Easy Laser Printer Maintenance and Repair

Stephen J. Bigelow

Windcrest®/McGraw-Hill

New York San Francisco Washington, D.C. Auckland Bogotá
Caracas Lisbon London Madrid Mexico City Milan
Montreal New Delhi San Juan Singapore
Sydney Tokyo Toronto

pbk   9 10 DOC/DOC 9 9 8
hc   1 2 3 4 5 6 7 8 9 DOC/DOC 9 9 8 7 6 5 4

Library of Congress Cataloging-in-Publication Data
Bigelow, Stephen J.
    Easy laser printer maintenance and repair / by Stephen J. Bigelow.
      p.   cm.
    Includes index.
    ISBN 0-07-035975-X   ISBN 0-07-035976-8 (pbk.)
    1. Laser printers—Maintenance and repair. I. Title.
TK7887.7.B538   1995
681'.62—dc20   94-33620
                                                       CIP

Acquisitions editor: Roland Phelps
Editorial team: B.J. Peterson, Editor
               Susan W. Kagey, Managing Editor
               Joanne Slike, Executive Editor
               Joann Woy, Indexer
Production team: Katherine G. Brown, Director
               Rhonda E. Baker, Coding
               Susan E. Hansford, Coding
               Brenda S. Wilhide, Computer Artist
               Lisa M. Mellott, Desktop Operator   0359768
Designer: Jaclyn J. Boone   EL1

Disclaimer and cautions

It is *important* that you read and understand the following information. Please read it carefully!

The repair of laser printers involves some amount of risk to yourself and other people in your work area. Use *extreme* caution when working with laser beams, ac (alternating current), and high-voltage power sources. Every reasonable effort has been made to identify and reduce areas of personal risk. Read this book carefully *before* attempting the procedures discussed. If you are uncomfortable following the procedures that are outlined in this book, do *not* attempt them—refer your service to qualified service personnel.

Neither the author, the publisher, nor anyone directly or indirectly connected with the publication of this book shall make any warranty EITHER EXPRESSED OR IMPLIED, with regard to this material, including, but not limited to, the implied warranties of QUALITY, merchantability, and fitness for ANY particular purpose. Further, neither the author, publisher, nor anyone directly or indirectly connected with the publication of this book shall be liable for errors or omissions contained herein, or for incidental or consequential damages, injuries, or financial or material losses resulting from the use, or inability to use, the material contained herein. This material is provided AS-IS, and the reader bears all responsibilities connected with its use.

Dedication

This book is dedicated to my wonderful wife, Kathleen.
Without her loving support and encouragement,
this book would still have been possible, but not nearly worth the effort.

Contents

APPENDICES

Acknowledgments

In today's fast-moving world of personal computers, it is virtually impossible to prepare a thorough, complete book on computer service without the cooperation and support of other individuals and corporations. I offer my heartfelt thanks to the following for material and technical support that helped to make this book:

Mr. Jim Broich, Hewlett-Packard Co.
Mr. Gregg Elmore, B+K Precision
Mr. Reza Ghalebi, ROHM Corp.
Ms. Jan Marciano, Epson America Corp.
Mr. Ron Trumbla, Tandy Corp.
Ms. Kirsten Wisdom, Hewlett-Packard Co.

Of course, I also wish to thank the editors and the development staff at McGraw-Hill for their endless patience and consideration in making this book a reality.

Introduction

EP (electrophotographic) printers have become an invaluable tool in the high-tech revolution. It seems that about every computer owner has an EP printer on hand or has access to one at work. You can find EP printers in offices, laboratories, stores, classrooms, factories, and homes throughout much of the world. Since EP printers became widely available in the early 1980s, they have rapidly evolved into flexible, reliable, and surprisingly cost-effective devices using any one of several mature printing technologies. This book discusses those technologies in detail and shows you how many of today's EP printers work.

No matter how reliable a laser or LED printer might be, it will eventually need some sort of maintenance and repair. This book is intended to provide the needed background information on printer mechanics and electronics, along with the techniques that guide you through the troubleshooting and repair of about any commercial EP printer. The book is written for the hobbyist or repair technician who has an intermediate knowledge of electronics and some working knowledge of mechanics. A knowledge of elementary troubleshooting is helpful, but it is not a prerequisite—this book describes the basic operation and use of several common test instruments. Troubleshooting procedures are presented in a discussion format that aids your overall understanding of the EP printer and how it works.

Easy Laser Printer Maintenance and Repair begins slowly with a detailed explanation of the *electrophotographic* concept. Chapter 1 shows you how to interpret the sometimes-confusing features and specifications of EP printers. A brief discussion of typical assemblies explains the purpose and relationship of each major printer subsection. Chapter 1 is a handy introduction for readers who are new to EP printers. Chapter 2 offers a review of the many diverse components that you will find in a typical EP printer. Mechanical, electromechanical, passive, and active components are covered in detail. Even experienced readers will find chapter 2 to be a helpful review. Chapter 3 presents detailed information on the practical tools and test equipment needed to perform many detailed repairs.

Hand tools, soldering and desoldering, multimeters, logic probes, and oscilloscopes are discussed carefully. Even if you've never held a soldering iron or test instrument before, this section shows you how to use each device. Service guidelines are covered next in chapter 4. You will learn the universal troubleshooting cycle, and understand the hazards of ac and static electricity.

Once you have the preliminary chapters out of the way, use chapter 5 to dig in to the theory behind electrophotographic concepts. You can learn each step in the EP process and how each contributes to creating the final paper image. You can make a comparison of laser, *LED* (light-emitting diode), and *LSC* (liquid crystal shutter) mechanisms. The real troubleshooting begins in chapter 6 with a thorough evaluation of linear power supplies, followed by an equally complete review of switching power supplies. Troubleshooting procedures are presented for each section. High-voltage supplies and troubleshooting are also covered.

Chapter 7 details each element of EP image formation. Each step along the way offers comprehensive troubleshooting procedures. Chapter 8 covers the operations and troubleshooting of the mechanical systems of a printer, including paper handling, sensors and interlocks, laser scanners, and EP cartridges. Chapter 9 provides a comprehensive discussion of the printer *ECP* (electronic control package). You can learn to understand and troubleshoot communication, memory, controls, and main logic.

Appendix A provides you with a reliable cleaning and maintenance procedure. Each of the troubleshooting symptoms and procedures in the book are condensed into a quick-reference form in appendix B. Look to appendix C for a detailed list of parts, materials, and service vendors.

Following the appendices is a comprehensive EP printer glossary.

I am interested in your success! I have taken every possible measure to ensure that this book is thorough and comprehensive. Your comments, questions, and suggestions about this book, as well as any of your personal troubleshooting experiences that you might like to share, are welcome at any time. Feel free to contact me directly:

Stephen J. Bigelow
Dynamic Learning Systems
P.O. Box 805
Marlboro, Massachusetts 01752
U.S.A.

CIS:
73652, 3205

1
CHAPTER

The electrophotographic printer

Laser printers (Fig. 1-1) have come a long way in the last few years. Not too long ago, laser printers were expensive commodities restricted to the few serious businesses that could afford them. However, the precision, speed, printing consistency, and image quality offered by laser printers made them extremely attractive peripherals. As computer designs continued to advance into the 1990s, laser printers also advanced while printer prices plummeted. Today, laser printers are available for under $600 (U.S.) and have become commonplace in homes and small businesses all over the

1-1 A Hewlett-Packard LaserJet III printer.

Hewlett-Packard Co.

world. This book shows you how modern laser printers work and explains how to maintain them effectively.

An overview

The intense competition between printer manufacturers has resulted in a staggering variety of laser printer models—each with different sizes, shapes, and features. In spite of this physical diversity, every laser printer ever made performs the same set of functions to transcribe the output of a computer into some permanent paper form. The process seems simple enough, right? In reality, however, it requires a complex interaction of electrical, electronic, and mechanical parts all working together to make a practical laser printer. Stop for a moment and consider some things that a laser printer must be capable of.

First, the laser printer can do nothing at all without a host computer to provide data and control signals, so a communication link must be established. To operate with any computer system, the printer must be compatible with one or more standard communication interfaces that have been developed. A printer must be able to use a wide variety of paper types and thicknesses, which can include such things as envelopes and labels. It must be capable of printing a vast selection of type styles and sizes, as well as graphics images, then mix those images together onto the same page.

The laser printer must be fast. It must communicate, process, and print information as quickly as possible. Laser printers must also be easy to use. Many features and options are accessible with a few careful strokes of the control panel. Paper input and output must be convenient. Expendable supplies such as toner should be quick and easy to change. Finally, laser printers must be reliable. They must produce even and consistent print over a long working life—often more than 300,000 pages (expendable items must be replaced more frequently).

What is electrophotographic?

You are probably wondering why people use the term *electrophotographic printer* when talking about laser printers. In truth, electrophotographic (or *EP*) is a broad term that refers to a printer that functions using the electrophotographic process. Laser printers are electrophotographic printers that use a laser beam to write image data, but there are also LED (light-emitting diode) page printers that use a bar of microscopic LEDs, instead of a laser beam, to write image data. Both laser and LED printers are electrophotographic printers (although laser-type printers are more common). You can learn much more about the electrophotographic process and see how laser and LED printers work in this book. In this book, the terms *EP printer*, *laser printer*, and *LED printer* are interchangeable.

Features and specifications

Make it a point to know your laser printer specifications and features before you begin any repair. The specifications and features give you a good idea what the printer can do, which might help you to test it more thoroughly during and after your repair. A listing of specifications is usually contained in an introductory section of the printer instruction manual or at the end in an appendix. If you do not have a copy of the printer documentation on hand, the manufacturer can often fax a copy of the specifications directly to you. Remember that there is no standard format for listing printer specifications. The format is up to the preferences of each manufacturer. Regardless of how the specifications are listed, you will most often find the following subjects: power requirements, interface compatibility, print capacity, print characteristics, reliability/life information, environmental information, and physical information. Each of these specifications has some importance, so you should be familiar with them in detail.

Power requirements

As with any electrical device, a printer requires power to function. Voltage, frequency, and power consumption are the three typical specifications that you will find here. Domestic U.S. voltage can vary from 105 to 130 Vac (alternating-current volts) at a frequency of 60 Hz (hertz). European voltage can range from 210 to 240 Vac at 50 Hz. Many current laser printers have a power selection switch that toggles the printer between 120 and 240 V operation. Power consumption is rated in *watts* (W). Depending on the particular model, laser printers can use up to 900 W during printing. However, most models use an automatic power-down mode that shuts down the major power-consuming components after the printer is idle for several minutes. Chapter 6 discusses the operation and repair of laser printer power supplies.

Interface compatibility

A printer is a *peripheral* device. That is, it serves no purpose at all unless it can communicate (or *interface*) with a computer. A communication link between printer and computer can be established in many different ways, but three interface techniques have become standard: RS-232, Centronics, and IEEE 488. Only a properly wired and terminated cable is needed to connect the printer and computer. Printer communication and troubleshooting are discussed in chapter 9.

RS-232 is a serial interface used to pass binary digits (or *bits*) one at a time between the computer and printer. Serial links of this type are very common, not just for printers, but for other serial communication applications such as modems and simple digital networks. RS-232 is popular due to its high speed, physical simplicity, and its ability to handle data over long distances.

Centronics is the standard for parallel communication. Although one used exclusively by printers, parallel communication has become popular for other peripherals such as parallel-port tape drives and portable CD-ROM (compact disc read-only memory) drives. Centronics is a de facto standard, so it is not officially endorsed by standards organizations such as the IEEE (Institute of Electrical and Electronic Engineers), EIA (Electronic Industries Association), or CCITT (International Consultative Committee for Telephone and Telegraph). Instead of passing one bit at a time, Centronics interfaces pass entire characters from the computer to the printer as sets of bits. Centronics is popular because of its functional simplicity. Although parallel connections require more interconnecting signal wires than an RS-232 cable, the hardware required to handle parallel information is simpler.

IEEE 488, also known as GPIB (general-purpose interface bus), is an official IEEE standard for parallel communication. It is not as widely used as Centronics or RS-232, but GPIB supports network and bidirectional communication between instruments. The GPIB technique was originally developed by Hewlett-Packard Company, where it is still widely used in their line of printers and plotters.

Print capacity

Print capacity is a generic term including several different laser printer specifications that outline what a printer can do. One of the most common print capacity specifications is *print speed*, which is measured in pages per minute (ppm). Inexpensive EP printers work at 4 ppm, but 8 to 10 ppm printers are available. Next, you must be concerned with *resolution*, which is the number of individual dots that can be placed per linear inch. Typical EP printers offer 300 × 300 dpi (dots per inch) resolution (300 lines per inch at 300 dots per linear inch, or 90,000 dots per square inch). A resolution this high is adequate for most business and personal graphics. The current generation of laser and LED printers is capable of 600 × 600 dpi resolution.

You might find a section on *paper specifications*. Although dot matrix and ink jet printers are very flexible in accepting a wide variety of paper thicknesses and finishes, the paper used in laser printers must fall within certain weights and finishes if the EP process is to work correctly. In most cases, standard letter-size, xerography-grade paper (16–24 pound bond) will work. The paper also should have a plain finish. Shiny or gloss-finished papers will cause problems with the EP process. Most laser printers will handle envelopes, transparencies, and labels. Before choosing such materials, be certain that they are labeled as safe or tested with laser printers. Poor-quality materials can jam and damage your printer.

Memory is another important specification for EP printers. Because laser printers assemble images as full pages of individual bits, the more memory that is available, the larger and more complex an image can be printed. Typical laser printers offer 512 K (kilobytes) to 1 Mb (megabytes), but 2–3 Mb is required for full-page graphic images that might be produced with software such as CorelDraw. Most laser printers offer memory upgrade options.

Print characteristics

Print characteristics specify just how printer images will appear, how they will be produced, or how characters from the printer will be interpreted. Fonts, software emulation, and character sets are the three specifications that you should be most familiar with.

A *font* is a style of type with certain visual characteristics that distinguish it from other type styles. These characteristics might include differences in basic character formation, accents, and decorative additions (that is, Courier versus Helvetica type). Figure 1-2 is an example of several basic printer fonts. Early laser printers relied on font cartridges that contained ROMs (read-only memories) that held the image data for each font. To change a font, you changed the cartridge. However, with the rise of Microsoft Windows and improved memory systems, most current laser printers use

10 point Arial

10 **POINT ALGERIAN**

12 **point Britannic Bold**

12 point Century Gothic

14 point Impact

16 POINT STENCIL

16 point Times New Roman

18 point Vivaldi

18 point Wide Latin

1-2 A sample of software-driven EP printer fonts.

soft fonts, where the data for desired fonts is downloaded to the printer during the actual printing process. Soft fonts also allow easy enhancements such as underlining, bold, italic, superscript, subscript, and so on.

All printers use their own built-in software "language" that is in the printer permanent memory. The language specifies such things as font formation (dot placement) and size, how to recognize and respond to control codes or control panel input, and more. This software language also tells the printer how to operate, communicate, and respond to problems.

Most of these languages were originally developed by leading printer manufacturers such as Hewlett-Packard, IBM, and Epson. Other manufacturers that wish to make their printers functionally compatible must use a software language that *emulates* one or more of the existing language standards. For example, most laser printers will emulate the operation of a Hewlett-Packard LaserJet III. *Emulation* in this use means that even though a laser printer is physically and electronically different from a LaserJet III, it will respond as if it were a LaserJet III when connected to a host computer. Another typical printer language is Postscript.

Ordinarily when a character code is sent to a printer, it is processed and printed as a fully formed alphanumeric character or other special symbol. However, because a character code is not large enough to carry every possible type of text or special symbol (for example, foreign-language characters or block graphics), characters are grouped into *character sets* that the printer can switch between. Switching a character set is often accomplished through a series of computer codes or control panel commands. A standard character set consists of 96 ASCII (American Standard Code for Information Interchange) characters. The 96 characters include 26 uppercase letters, 26 lowercase letters, 10 digits, punctuation, symbols, and some control codes. Other character sets can include 96 italic ASCII characters, international characters (German, French, Spanish, etc.), and unique block graphics.

Reliability/life information

Reliability and *life expectancy* information expresses the expected working life of the laser printer or its components in pages or time. For example, a typical EP toner cartridge is rated for 200–250 pages, and the image-formation "engine" is rated for up to 300,000 pages. Many printers are rated at 5,000 pages per month (about 200 pages per workday). You might see this same information expressed as MTBF (mean time between failures).

Environmental information

Environmental specifications indicate the physical operating ranges of your printer. *Storage temperature* and *operating temperature* are the two most common environmental conditions. A typical laser printer can be stored in temperatures between –10 and 50°C, but can only be used from 10 to 32°C (on average). It is a good idea to let your printer stabilize at the ambient temperature and humidity for several hours before operating it. *Relative humidity* can often be allowed to range from 10 to 90% during storage, but must be limited to a range of 40 to 70% during operation. Keep in mind that humidity limits are given as noncondensing values. *Noncondensing* means that you cannot allow water vapor to condense into liquid form. Liquid water in the laser printer would certainly damage its image-formation system.

Your printer also might specify *physical shock* or *vibration* limits to indicate the amount of abuse the printer can sustain before damage can occur. Shock or vibration is usually rated in units of *g-force*. Keep in mind that laser printers are remarkably delicate devices; any substantial shock or vibration might disturb the optics that direct the laser beam. LED printers are a bit more rugged, but also use optics that can be damaged or misaligned by rough handling.

Physical information

Physical information about a printer includes such routine data as the printer height, width, depth, and weight. In some cases, an operating noise level specification is included to indicate just how loud the printer will be during operation and standby. Noise specifications are usually given in dBA (A-weighted decibels).

Typical assemblies

No matter how diverse or unique EP printers might appear from one model to another, their differences are primarily cosmetic. It is true that each printer might use different individual components, but every laser printer must perform a very similar set of actions. As a result, most laser printers can be broken down into a series of typical sections, or functional areas as shown in Fig. 1-3. Before you troubleshoot a printer, you must understand the purpose of each area.

The ac power supply

An ac power supply is usually a simple electronic module that provides energy for the fusing assembly heaters and erase lamp assembly. There is typically little that goes wrong with the ac supply unless a serious fault in the fuser or erase assemblies damages the supply. You can learn about power supply operation and repair in chapter 6.

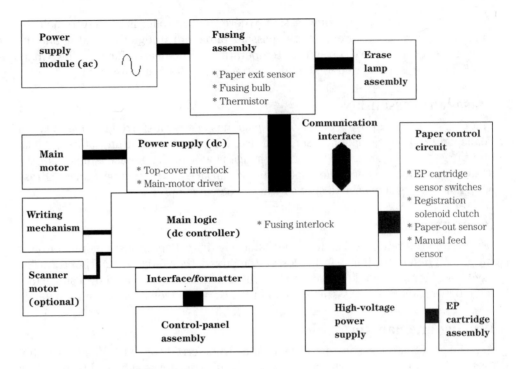

1-3 Diagram of a typical laser printer.

The dc power supply

The dc (direct current) power supply converts main ac entering the printer into one or more levels of dc that are used to power the printer electronics and electro-mechanical devices (such as motors and clutches). Like the ac supply, a dc power supply is a fairly rugged and reliable device unless a fault in some part of the printer circuitry damages the supply. A cover interlock in the dc supply shuts down printer operation if a protective cover is opened. You can learn about power supply operation and repair in chapter 6.

High-voltage power supply

The EP process relies on high-voltage (1,000 V or more) that is used to create and dissipate the powerful static charges that move toner within an EP printer. Even though specialized components are used in high-voltage supplies, high voltages place a great deal of stress on electronic parts, so high-voltage supplies tend to fail more commonly than ordinary ac or dc supply modules. See chapter 6 for more information on power supplies.

Fusing assembly

Images are developed on paper using a fine powder called *toner*. The toner must be fixed to the paper (otherwise, the toner would smudge or simply blow off the page). Heat and pressure are used to *fuse* toner to the paper. The fusing assembly uses a set of two rollers in compression, where the top roller is heated to melt the toner. A paper exit sensor detects the passage of paper through the printer, and the thermistor sensor is used to regulate temperature in the heated fusing roller. Chapters 5 and 7 present more detailed information about the fusing assembly.

Erase lamp assembly

The image that appears on a printed page has been transferred there as a latent image written to a special photosensitive drum. Each time the drum rotates, the latent image must be erased before a new image is written. The erase lamps clear the drum thoroughly and allow the photosensitive surface to accept a new image. Erase lamp failure is usually easy to spot as you can see in chapter 7.

Main motor

EP printers rely on substantial mechanical activity. Paper must be drawn from a supply tray, fed to the image formation system, fixed, then fed to the output tray. The mechanical force needed to support all of these activities is provided by a single motor and mechanical drive assembly. Chapter 8 describes mechanical systems in detail.

Writing mechanism

The data that makes up an image must be transferred (or "written") to the photosensitive drum. As you can see in chapter 5, this transfer is achieved by directing light across the drum surface. For a laser printer, writing is accomplished by scanning a laser beam across a drum. For an LED printer, the light generated by individual microscopic LEDs (one LED for every dot) transfers image data to the drum. Writing is controlled by the main logic assembly (or ECP).

Scanner-motor assembly

When a laser is used as a writing mechanism, the beam must be scanned back and forth across the drum surface. This scanning process uses a hexagonal mirror that is rotated with a motor. Note that scanners are not needed for LED printers because there is no beam to scan across the drum. You can find more information on scanner assemblies in chapter 8.

Paper-control assembly

Paper must be grabbed from the paper tray, registered with the latent image, passed through the image formation system, fused, and passed out of the printer. Although the main motor turns constantly, not all portions of the paper handling system can be in motion at all times. The paper control assembly provides the sensors that detect the presence of paper in the paper tray, the presence of paper in the manual feed slot, and the sensitivity of each EP cartridge for optimum printing. In addition to sensors, the paper-control assembly provides the paper pickup and registration roller clutches that grab and register the page during printing. Chapter 8 discusses the paper-handling assembly in detail.

Main logic assembly

The main logic assembly (or electronic control package—ECP) is the heart and soul of your EP printer. The main logic assembly has most of the circuitry that operates the printer, including electronics that communicates with the computer and control panel. Main logic is also responsible for checking and responding to input provided by a variety of sensors. Problems that occur in main logic can range from subtle problems to major malfunctions. Chapter 9 covers the main logic circuits found in EP printers.

EP cartridge assembly

The Electrophotographic (EP) cartridge of a printer is a remarkable piece of engineering that combines the toner supply and much of the printer image-formation system into a single, replaceable cartridge. By replacing the EP cartridge, you also replace delicate, wear-prone parts such as the primary corona, EP drum, and developer roller. The modularity of the cartridge simplifies the maintenance of your printer and improves its overall reliability. The image-formation system is discussed in chapter 7.

Control-panel assembly

Users must be able to interface with the printer to select various options or operating modes. Not only do current control panels provide multifunction buttons, but most also provide an LCD display for printer status and menu prompts. The control panel is covered with main logic in chapter 9.

2
CHAPTER

Typical components

To troubleshoot any electromechanical system, you must be familiar with the individual electronic and mechanical components that you find inside. This chapter introduces you to a cross-section of components found in most typical laser printers. Your troubleshooting efforts are simplified if you can identify important components on sight, understand their purpose, and spot any obvious defects. Keep in mind that this chapter is by no means a complete review of every possible type of component, but it will give you a good idea of what to expect.

Mechanical parts

For the purpose of this discussion, mechanical parts basically serve a single purpose—to transfer force from one point to another. For example, a laser printer uses a single motor to operate the printer paper transport and image-formation systems. The physical force of a motor must be transferred to the paper, as well as to the various rollers and mechanisms that make the printer work. This transfer is accomplished through a series of gears, pulleys, rollers, and belts. Whenever mechanical parts are in contact with one another, they produce friction that causes wear. Lubricants, bushings, and bearings work to minimize the damaging effects of friction. Chapter 8 presents troubleshooting procedures for mechanical systems.

Gears

Gears perform several important tasks. Their most common application is to transfer mechanical force from one rotating shaft to another. The simplest arrangement uses two gears in tandem as in Fig. 2-1. When two gears are used, the direction of *secondary* rotation is opposite that of the *primary* shaft. If secondary direction must be the same, a third gear can be added as shown in Fig. 2-2. The orientation of applied force can be changed by using angled (or *beveled*) gears shown in Fig. 2-3. By varying the angles of both gears, force can be directed almost anywhere. Several secondary gears can be run from a single drive gear to distribute force to multiple locations simultaneously, which is a critical feature for laser printer operation.

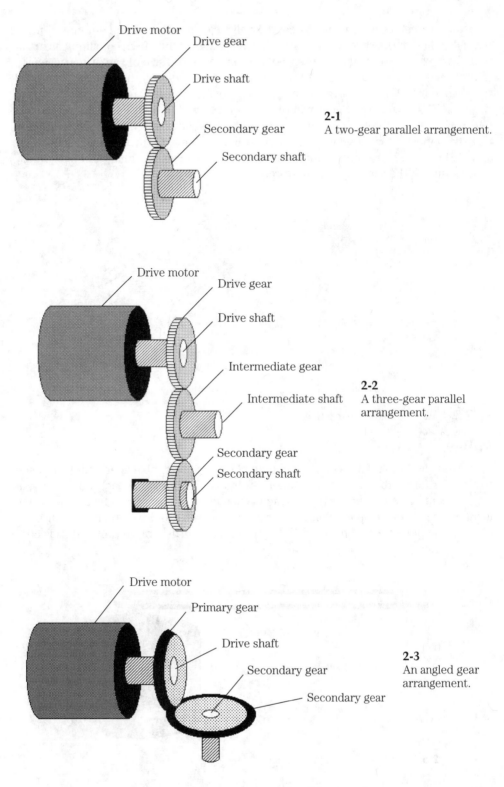

Drive motor
Drive gear
Drive shaft
Secondary gear
Secondary shaft

2-1
A two-gear parallel arrangement.

Drive motor
Drive gear
Drive shaft
Intermediate gear
Intermediate shaft
Secondary gear
Secondary shaft

2-2
A three-gear parallel
arrangement.

Drive motor
Primary gear
Drive shaft
Secondary gear
Secondary gear

2-3
An angled gear
arrangement.

Not only can gears transfer force, they also can alter speed and amount of force that is applied at the secondary shaft. Figure 2-4 shows the effects of simple gear ratios. *Gear ratios* are usually expressed as the ratio of the size of the primary gear to that of the secondary gear (or of the number of gear teeth on the primary to that of the secondary). For a *high* ratio, the primary gear is larger than the secondary gear. As a result, the secondary gear will turn faster, but with less force. (For mechanical parts that rotate, the force in the rotating parts is known as *torque*). The effect is just the opposite for a *low* ratio. A small primary gear will turn a larger secondary slower, but with more force. Finally, an equal ratio causes a primary and secondary gear to turn at the same speed and force.

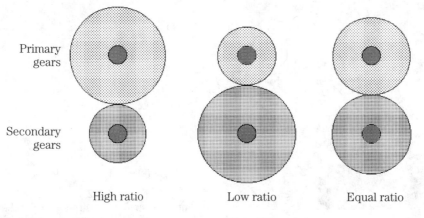

Primary gears

Secondary gears

High ratio   Low ratio   Equal ratio

2-4 An example of gear ratios.

Pulleys

Pulley assemblies are common in many laser printers. Like gears, they are used to transfer force from one point to another. Instead of direct contact, however, pulleys are joined by a *drive linkage*, which is usually a belt, wire, or chain. The action is much the same as the fan belt in your automobile or the drive chain on your bicycle. A basic pulley set is shown in Fig. 2-5. A motor turns a drive pulley that is con-

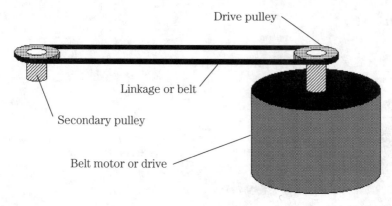

Drive pulley

Linkage or belt

Secondary pulley

Belt motor or drive

2-5 Top view of a basic pulley set.

nected to a secondary pulley through a drive belt that is under tension. As the drive pulley turns, force is transferred to the secondary through the linkage, so the secondary pulley also turns. Notice that both pulleys turn in the same direction. A pulley/belt configuration is sometimes used as a conveyor belt to carry paper evenly to the fusing rollers.

Pulleys and drive linkages will vary depending on their particular application. Low-force applications can use narrow pulleys (little more than a wheel with a groove in it) connected with a wire linkage. Wire is not very rugged, and its contact surface area with both pulleys is small. Therefore, wire can slide when it stretches under tension or if load becomes excessive. Belts and their pulleys are wider, so there is much more surface contact around each pulley. Belts are usually stronger than wire, so there is less tendency to stretch under tension. Greater strength makes belt-driven pulleys better suited for heavier loads.

You can replace pulleys with sprocket wheels and a chain linkage. Because each chain link meshes with the sprocket wheels, any chance of slipping is eliminated. Chains are almost immune to stretching under tension, so chain drives are used to handle the highest loads.

Rollers

Rollers are really a focal point of laser printer operation. Rollers not only serve to grab a sheet of paper from the paper tray, but rollers position (or *register*) the paper before printing. The image formation system uses several rollers to distribute toner and transfer the image to paper. By passing the paper through a set of heated rollers, the transferred image is fused to paper. Damaged, old, or dirty rollers might have an adverse effect not only on paper handling, but on overall image quality.

Reducing friction

As with all mechanical systems, parts that are in contact with one another will wear while the system operates due to unavoidable friction that occurs between parts. Therefore, reducing friction will extend the working life of your printer. Lubrication, bushings, and bearings are three commonly accepted methods of reducing friction.

Use of oils or grease is one way to reduce friction (and might prove effective in small doses), but this *lubrication* must be replaced regularly for it to remain effective. Otherwise, it can wear away, dry out, or harden into thick sludge. Lubricants also are notorious for collecting dust and debris from the environment, which eventually defeats any benefits that the lubricant can provide.

Bushings are usually "throw-away" wear surfaces as shown in Fig. 2-6. A bushing is made of softer materials than the parts it is separating, so any friction generated by moving parts will wear out the bushing before the other parts touch each other. When a bushing wears out, simply replace it with a new one. Bushings are much less expensive and easier to replace than major mechanical parts such as slides or frames. Today, bushing materials are reliable, and can last throughout the working life of the printer.

Probably the most effective devices for reducing friction between parts are *bearings*. Bearings consist of a hard metal case with steel balls or rollers packed inside as shown in Fig. 2-7. Because each steel ball contacts a load-bearing surface at

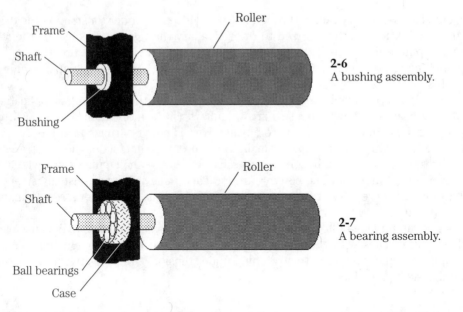

2-6
A bushing assembly.

2-7
A bearing assembly.

only one point, friction (and wear) is substantially lower than for bushings. Unfortunately, bearing assemblies are often much more expensive than bushings, so bearings are used only to handle heavy loads, or in places that would be too difficult to change bushings. Most laser printers avoid the expense of bearings in favor of inexpensive bushings.

Electromechanical components

Electromechanical components are a particular class of devices that convert electrical energy into mechanical force or rotation. Relays, solenoids, and motors are three common electromechanical components that you should understand. Each of these important devices relies on the principles of *electromagnetism*.

Electromagnetism

Whenever electrical current passes through a conductor, a magnetic field is generated around the circumference of that conductor as shown in Fig. 2-8. Such a magnetic field can exert a physical force on *permeable* materials (any materials that can be magnetized). The strength of a magnetic field around a conductor is proportional

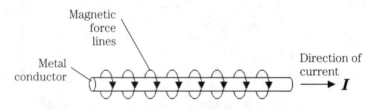

2-8 The magnetic field formed around a conductor.

to the amount of current flowing through it. Higher amounts of current result in stronger magnetic fields, and vice versa.

Unfortunately, it is virtually impossible to pass enough current through a typical wire to produce a magnetic field that is strong enough to do any useful work. The magnetic field must somehow be concentrated, usually by *coiling* the wire as shown in Fig. 2-9. When arranged in this way, the coil takes on magnetic poles—just like a permanent magnet. Notice how the direction of magnetic flux always points to the *north* pole of the coil. If the direction of current flow were reversed, the magnetic poles of the coil also would be reversed.

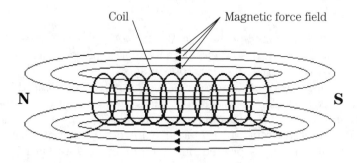

2-9 The concentration of magnetic force in a coil.

To concentrate magnetic forces even further, a permeable core material can be inserted into the coil center as in Fig. 2-10. Typically, iron, steel, and cobalt are considered the classical core materials, but iron-ceramic blends are used as well. Coils of wire such as these form the foundation of all electromechanical devices.

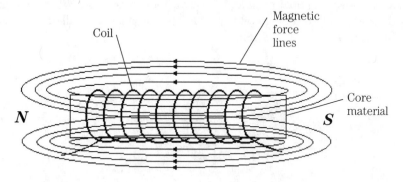

2-10 Concentrating magnetic force with a permeable core.

Relays

A *relay* is simply a mechanical switch that is actuated with the electromagnetic force generated by an energized coil. A diagram of a typical relay is shown in Fig. 2-11. The switch (or *contact set*) can be normally open (NO) or normally closed (NC) while the coil is de-energized. When activated, the magnetic field of the coil causes normally open contacts to close, or normally closed contacts to open. Con-

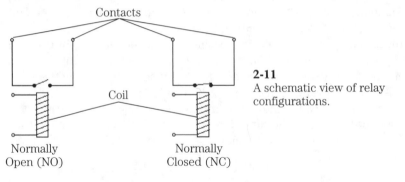

2-11
A schematic view of relay configurations.

tacts are held in their actuated positions as long as the coil is energized. If the coil is turned off, contacts will return to their normally open or closed states. Keep in mind that a coil might drive more than one set of contacts.

Relays are not always easy to recognize on sight. Most relays used in electronic circuits are housed in small rectangular containers of metal or plastic. Low-power relays are available in oversized IC (integrated circuit) style packages and soldered right into a PC (printed circuit) board just like any other integrated circuit. Unless the relay internal diagram is printed on its outer case, you will need a printer schematic or manufacturer's data for the relay to determine the proper input and output functions of each relay pin.

Solenoids

Solenoids convert electromagnetic force directly into motion as shown in Fig. 2-12. Unlike ordinary electromagnets whose cores remain fixed within a coil, a solenoid core is allowed to float back and forth without restriction. When energized, the magnetic field generated by a coil exerts a force on its core (called a *plunger*)that pushes it out from its rest position.

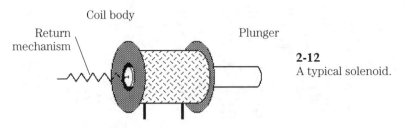

2-12
A typical solenoid.

If left unrestrained, a plunger would simply shoot out if its coil and fall away. Plungers are usually retained by a spring or some other sort of mechanical return assembly. Being retained, a plunger extends only to some known distance when the coil is fired; then it automatically returns to its rest position when the coil is off.

Solenoids are commonly used as clutches in laser printer paper-handling systems. Remember that the motor providing mechanical torque in the printer is constantly running while a page is being printed. However, the paper-grab and registration rollers need only turn briefly during the printing cycle. Solenoid "clutches" are used to engage and disengage the motor torque as needed.

Motors

Motors are an essential part of every laser printer manufactured today. Motors operate the entire mechanical transport system. Chapter 8 presents a detailed discussion of mechanical systems and service. For now, concentrate on the motor itself.

All motors convert electrical energy into rotating mechanical force (torque). In turn, that force can be distributed with mechanical parts to turn a roller or move a belt. An *induction* motor provides torque through a series of powerful electromagnets (coils) around a permanent magnet core as shown in Fig. 2-13. The core (known as a *rotor*) is little more than a shaft that is free to rotate as its poles encounter electromagnetic forces. Each coil (also called a *phase* or *phase winding*) is built into the motor stationary frame (or *stator*).

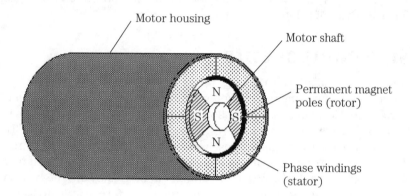

2-13 Simplified view of a typical induction motor.

By powering each phase in its proper order, the rotor can be made to turn with some amount of force. The amount of angular force generated by a motor is known as *torque*. Induction motors generally require two ac signals separated by a 90 degree phase difference. These sinusoidal driving signals vary the strength of each phase evenly to create smooth rotation. Induction motors are rarely used in today's commercial printers because they do not lend themselves to the precise positioning requirements of most printers. Instead, a close cousin of the induction motor is used, and the motor is called a *dc servo* motor.

The dc servos are powered by bipolar dc signals rather than by sinusoidal ac. Like the induction motor, a dc servo is turned by bipolar driver signals in rapid succession. Although dc servos are not very precise for positioning, they can hold a constant speed very accurately. This characteristic makes dc servos ideal for driving the scanner mirrors in laser printers that must rotate at a precise rate.

A popular variation of the dc servo motor is called a *stepping motor*. Physically, a stepping motor (or *stepper*) is very similar to a dc servo, but the rotor/stator arrangement is much more intricate. Like dc servos, stepper motors are driven by a series of square wave pulses separated by a phase difference. The sudden shift in drive signals coupled with the intricate mechanical arrangement causes the rotor to jump (or step) in certain angular increments, not a smooth, continuous rotation. Once the rotor has reached its next step, it will hold its position as long as driver signals maintain their conditions. If driver signals hold steady, the motor could remain stationary indefinitely. A typical stepping motor can achieve 1.8 degrees per step, which means a motor must make 200 (360 degrees divided by 1.8 degrees) individual steps to complete a single rotation. However, gear ratios can break down motor movement into much finer divisions.

Stepping motors are ideal for precise positioning. Because the motor moves in known angular steps, it can be rotated to any position simply by applying the appropriate series of driver pulses. For example, suppose your motor had to rotate 180 degrees. If each step equals 1.8 degrees, you need only send a series of 100 (180 degrees divided by 1.8 degrees) pulses to turn the rotor exactly that amount. Logic circuits in the printer generate each pulse, then driver circuits amplify those pulses into the high-power signals that actually operate the motor. Chapter 9 explains the operation and repair of printer electronics.

Passive components

If you intend to do any subassembly or component-level troubleshooting of laser printer systems, you must have an understanding of the various electronic components that are available. Most circuits contain both *active* and *passive* components working together. Passive components include resistors, capacitors, and inductors. They are called *passive* because their only purpose is to store or dissipate circuit energy.

Active components make up a broader group of semiconductor-based parts such as diodes, transistors, and all types of integrated circuits. They are referred to as *active* because each component uses circuit energy to perform a specific set of functions—they all do something. A component might be as simple as a rectifier or as complex as a microprocessor, but active parts are the key elements in modern electronic circuits. This part of the chapter shows you each general type of component, how they work, how to read their markings, and how they fail.

Resistors

All resistors ever made serve a single purpose—to dissipate power in a controlled fashion. Resistors appear in most circuits, but they are usually used for such things as voltage division, current limiting, etc. Resistors dissipate power by presenting a resistance to the flow of current. Wasted energy is then shed by the resistor as heat. In printer logic circuits, so little energy is wasted by resistors that virtually no temperature increase is detectable. In high-energy circuits such as power supplies, resistors can shed substantial amounts of heat. The basic unit of resistance is the *ohm*. The symbol for resistance is the Greek symbol omega (Ω). You will see resistance also presented as kilohms (kΩ or thousands of ohms) or megohms (MΩ or millions of ohms).

Carbon-film resistors, as shown in Fig. 2-14, have largely replaced carbon-composition resistors in most circuits requiring through-hole resistors. Instead of carbon filling, a very precise layer of carbon film is applied to a thin ceramic tube. The thickness of this coating affects the amount of resistance—thicker coatings yield lower levels of resistance, and vice versa. Metal leads are attached by caps at both ends,

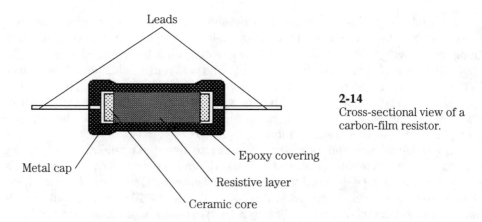

Leads

2-14
Cross-sectional view of a
carbon-film resistor.

Epoxy covering

Metal cap

Resistive layer

Ceramic core

and the entire finished assembly is dipped in epoxy or ceramic. Carbon-film resistors are generally more accurate than carbon composition resistors because a film can be deposited very precisely during manufacture.

A *surface-mount* resistor is shown in Fig. 2-15. As with carbon-film resistors, surface-mount resistors are formed by depositing a layer of carbon film onto a thin ceramic substrate. Metal tabs are attached at both ends of the wafer. Surface-mount resistors are soldered directly on the top or bottom sides of a printed circuit board instead of using leads to penetrate the PC board. Surface-mount resistors are incredibly small devices (only a few square millimeters in area), yet they offer very tight tolerances. Surface-mount resistors are used extensively in computers and printers.

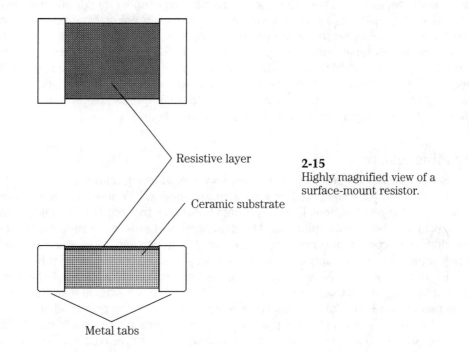

Resistive layer

2-15
Highly magnified view of a
surface-mount resistor.

Ceramic substrate

Metal tabs

Adjustable resistors, known as *potentiometers or rheostats*, are usually used in laser printers to adjust contrast by varying the level of high voltage. A typical potentiometer consists of a movable metal wiper resting on a layer of resistive film. Although the total resistance of the film, end-to-end, will remain unchanged, resistance between either end and the wiper blade will vary as the wiper is moved. There are two typical types of adjustable resistor: knob-type, where the wiper is turned clockwise or counterclockwise using a rotating metal shaft, or slide-type, where the wiper is moved back and forth in a straight line.

In addition to value and tolerance, resistors also are rated by their power-handling capacity. Power is normally measured in watts (W) and depends on the amount of current (I) and voltage (V) applied to the resistor as given by Ohm's law ($P = I \times V$). Resistors are typically manufactured in $\frac{1}{16}$, $\frac{1}{8}$, $\frac{1}{4}$, $\frac{1}{2}$, 1, 2, and 5 W sizes to handle a wide variety of power conditions. Size is directly related to power dissipation ability, so larger resistors usually can handle more power than a smaller resistor of the same value.

As long as power dissipation is below its rating, a resistor should hold its resistance value and perform indefinitely. However, when a resistor is forced to exceed its power rating, it cannot shed heat fast enough to maintain a stable temperature. Ultimately, the resistor will overheat and burn out. In all cases, a burned-out resistor forms an open circuit. A faulty resistor might appear slightly discolored, or it might appear burned and cracked. It really depends on the severity and duration of its overheating. Extreme overheating can burn a printed circuit board, and possibly damage the printed copper traces.

Failures among potentiometers usually take the form of intermittent connections between the wiper blade and resistive film. Remember that film slowly wears away as the wiper moves back and forth across it. Over time, enough film can wear away that the wiper cannot make good contact at certain points. The poor contact can cause all types of erratic or intermittent operation. With EP printers, it is rarely necessary to continually adjust printing contrast once optimum levels are found, so it is unlikely for adjustable resistors to wear out in laser printers, but dust and debris might collect and cause intermittent operation when adjustment is needed. Try cleaning an intermittent potentiometer with a high-quality electronic contact cleaner. Replace any intermittent potentiometers or rheostats.

Reading resistors

Every resistor is marked with its proper value. Marking allows resistors to be identified on sight and compared versus schematics or part layout drawings. Now that you know what resistors look like, you should know how to identify their value without having to rely on test equipment. There are three ways to mark a resistor: explicit marking, color coding, and numerical marking. It is important to decipher all three types of marks because many circuits use resistors with a mix of marking schemes.

Explicit marking is just as the name implies—the actual value of the part is written right on the part. Large, ceramic power resistors often use explicit marking. Their long, rectangular bodies are usually large enough to hold clearly printed characters.

Color coding has long been a popular marking scheme for carbon-film resistors that are simply too small to hold explicit markings. The twelve colors used in color coding are shown in Table 2-1. The first ten colors (black through white) are used as *value* and *multiplier* colors. Silver and gold colors serve as *tolerance* indicators.

Table 2-1. The standard resistor color code

Color	1st Band	2nd Band	Multiplier	Tolerance
Black	0	0	1	
Brown	1	1	10	
Red	2	2	100	
Orange	3	3	1,000	
Yellow	4	4	10,000	
Green	5	5	100,000	
Blue	6	6	1,000,000	
Violet	7	7	10,000,000	
Gray	8	8	100,000,000	
White	9	9	—	
(None)				± 20%
Silver				± 10%
Gold				± 5%

The color code approach uses a series of colored bands as shown in Fig. 2-16. Band number 1 is always located closest to the end of the resistor. Bands one and two are the value bands, and band three is the multiplier. A forth band (if present) will be silver or gold to indicate the resistor tolerance. On rare occasions, you might encounter a fifth band that indicates the reliability of a resistor (and is used only for military- and aerospace-grade resistors).

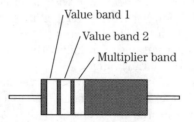

Value band 1
Value band 2
Multiplier band

2-16
Resistor color-coding scheme.

As an example of color coding, suppose the resistor of Fig. 2-16 offered a color sequence of brown, black, and red. Note from Table 2-1 that brown=1, black=0, and red=100 (because the red band occupies the *multiplier* position). The sequence would be read as [band 1][band 2] × [band 3] or 1 0 × 100, or 1,000 Ω (1 kΩ). If the first three color bands of a resistor read red, red, orange, the resistor would be read as 2 2 × 1,000, or 22,000 Ω (22 kΩ), and so on.

When a forth band is found, it shows the resistor tolerance. A gold band represents an excellent tolerance of ±5% of rated value. A silver band represents a fair tolerance of ±10%, and no tolerance band indicates a poor tolerance of ±20%. When a faulty resistor must be replaced, it should be replaced with a resistor of equal or smaller tolerance whenever possible.

Color-coded resistors are rapidly being replaced by surface-mount (SM) resistors. SM resistors are far too small for clear color coding. Instead, a three-digit numerical code is used (even though you might need a small magnifying glass to see it). Each digit corresponds to the first three bands of the color code as shown in Fig. 2-16. The first two numbers are *value* digits, and the third number is the *multiplier*. The multiplier digit indicates how many places to the *right* that the value's decimal place must be shifted. For the example of Fig. 2-17, a numerical code of 102 denotes a value of 10 with 2 zeros added on to make the number 1,000 (1 kΩ). A marking of 331 is read as 330 Ω, and so on.

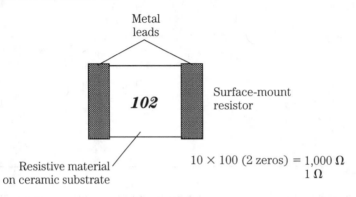

2-17 Surface-mount resistor markings.

Capacitors

Capacitors are simply energy-storage devices. They store energy in the form of an electrical charge. By themselves, capacitors have little practical use, but the capacitor principle has important applications when combined with other components in filters, resonant or timing circuits, and power supplies. Capacitance is measured in *farads* (F). In actual practice, a farad is a very large amount of capacitance, so most normal capacitors measure in the microfarad (μF or millionths of a farad) and picofarad (pF or millionths of a millionth of a farad) range.

In principle, a capacitor is little more than two conductive plates separated by an insulator (called a *dielectric*) as shown in Fig. 2-18. The amount of capacitance provided by this type of assembly depends on the area of each plate, their distance apart, and the dielectric material that separates them. Even larger values of capacitance can be created by rolling up a plate/dielectric assembly and housing it in a cylinder.

When voltage is applied to a capacitor, electrons will flow into it until it is fully charged. At that point, current stops flowing (even though voltage might still be applied), and voltage across the capacitor will equal its applied voltage. If applied voltage is removed, the capacitor will tend to retain the charge of electrons deposited on its plates. Just how long it can do this depends on the specific materials used to construct the capacitor, as well as its overall size. Internal resistance through the dielectric material will eventually bleed off any charge. For the purposes of this book, all you really need to remember is that capacitors are built to store electrical charge.

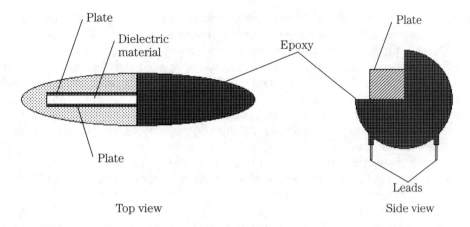

2-18 Sectional view of a conventional plate capacitor.

There are generally two types of capacitors that you should be familiar with. The types can be categorized as *fixed* or *electrolytic*. A selection of capacitor types is shown in Fig. 2-19. Fixed capacitors are nonpolarized devices—they can be inserted into a circuit regardless of their lead orientation. Many fixed capacitors are assembled as small wafers or disks. Each conductive plate is typically aluminum foil. Common dielectrics include paper, mica, and various ceramic materials. The complete assembly is then coated in a hard plastic, epoxy, or ceramic housing to keep out hu-

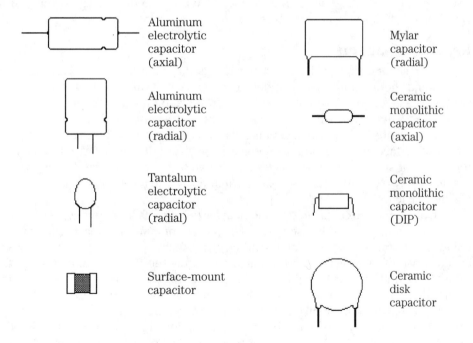

2-19 Outlines of various capacitor types.

midity. Larger capacitors can be assembled into large, hermetically sealed canisters. Fixed capacitors also are designed in surface-mount form.

Electrolytic capacitors are polarized components—they must be inserted into a circuit in the proper orientation with respect to the applied signal voltage. Tantalum capacitors are often found in a dipped (or *teardrop*) shape, or as small canisters. Aluminum electrolytic capacitors are usually used in general-purpose applications where polarized devices are needed. The difference between fixed and electrolytic capacitors is primarily in their materials, but the principles and purpose of capacitance remain the same.

Capacitors also are designated as *axial* or *radial* devices, which indicates the capacitor lead configuration. When both leads emerge from the same end of the capacitor, the device is said to be *radial*. If the leads emerge from either side, the capacitor is known as axial.

Surface-mount capacitors are usually fixed ceramic devices using a dielectric core capped by electrodes at both ends. If an electrolytic capacitor is needed, a surface-mount tantalum device is typically used. Although the construction of a surface-mount tantalum capacitor differs substantially from a ceramic surface-mount capacitor, they both appear very similar to the unaided eye. All polarized capacitors are marked with some type of polarity indicator.

Like resistors, most capacitors tend to be rugged and reliable devices. Because they only store energy (not dissipate it), it is virtually impossible to burn them out. You can damage or destroy capacitors if you exceed their working voltage (WV) rating or reverse the orientation of a polarized device. Damage can occur if a failure elsewhere in a circuit causes excessive energy to be applied across a capacitor, or if you should install a new electrolytic capacitor incorrectly.

Reading capacitors

Like resistors, all capacitors carry markings that identify their value. Once you understand the markings, you can determine capacitor values on sight. Capacitors are typically marked in two ways: explicit marking and numerical codes.

Explicit marking is used with capacitors that are physically large enough to carry their printed value. Large, ceramic disk, mylar, and electrolytic capacitors have plenty of surface area to hold readable markings. Note that all polarized capacitors, regardless of size, must show which of their two leads are positive or negative. Be certain to pay close attention to polarizer markings whenever you are testing or replacing capacitors.

Small, nonpolarized capacitors and many sizes of surface-mount capacitors now make use of numerical coding schemes. The pattern of numerical markings is easy to follow because it is very similar to the marking technique used with numerically coded resistors. A series of three numbers is used—the first two numbers are the value digits, and the third number is the multiplier digit (how many zeros are added to the value digits). Capacitor numerical marking is shown in Fig. 2-20.

Most capacitor numerical markings are based on picofarad measurements. Thus, the capacitor marked 150 would be read as a value of 15 with no zeros added (15 pF). A marking of 151 is 15 with one zero added (150 pF). The marking 152 would

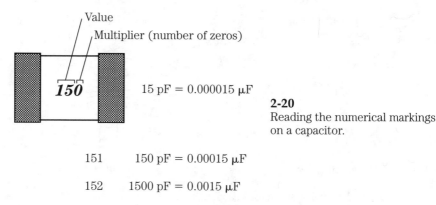

Value

Multiplier (number of zeros)

150

15 pF = 0.000015 µF

2-20
Reading the numerical markings
on a capacitor.

151	150 pF = 0.00015 µF
152	1500 pF = 0.0015 µF
153	15000 pF = 0.015 µF

be 15 with two zeros added (1,500 pF), and so on. A marking of 224 would be 22 with four zeros (220,000 pF). The decimal place is always shifted to the right.

Although this marking system is based on picofarads, every value can be expressed as microfarads (µF) simply by dividing the picofarad value by 1 million. For example, a 15 pF capacitor could be called a 0.000015 µF (15/1,000,000) capacitor. Of course, there is no advantage in marking such a small capacitor in the microfarad range when 15 pF is such a convenient value, but the conversion is a simple one. The 15,000 pF capacitor also could be shown as 0.015 µF (15,000/1,000,000). Capacitors with large picofarad values are often expressed more effectively as microfarads. To confirm your estimates, you can measure the capacitor with a capacitance meter.

Inductors

Like capacitors, inductors are energy-storage devices. Unlike capacitors, inductors store energy in the form of a magnetic field. Before the introduction of integrated circuits, inductors served a key role with capacitors in the formation of filters and resonant (tuned) circuits. Although advances in solid-state electronics have made inductors virtually obsolete in traditional applications, they remain invaluable for high-energy circuits such as power supplies. Inductors also are used in transformers, motors, and relays. Inductance is measured in *henrys* (H), but smaller inductors can be found in the millihenry (mH) or microhenry (µH) range.

Coils are available in many shapes and sizes as shown in Fig. 2-21. The particular size and shape will depend on the amount of energy that must be stored, and the magnetic characteristics desired. Laser printers use coils in their ac, dc, or high-voltage power supplies. Small coils also are available in surface-mount or leaded packages.

A *transformer* is actually a combination of inductors working in tandem. As Fig. 2-22 shows, a transformer is composed of three important elements: a *primary* (or input) winding, a *secondary* (or output) winding, and a *core* structure of some type. Transformers are used to alter (or transform) ac voltage and current levels in a circuit, as well as to isolate one circuit from another. An ac signal is applied to the primary winding. Because the magnitude of this input signal is constantly changing, the magnetic field it generates will constantly fluctuate as well. When this fluctuating

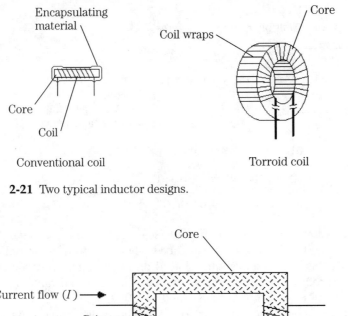

2-21 Two typical inductor designs.

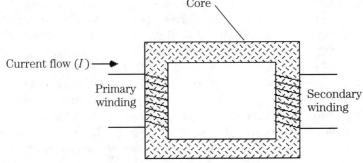

2-22 Diagram of a transformer.

field intersects the secondary coil, another ac signal is created (or *induced*) across it. This principle is known as *magnetic coupling*. Any secondary ac signal will duplicate the original signal. Primary and secondary windings are often wound around the same core structure that provides a common structure and efficient magnetic coupling from primary to secondary.

The actual amount of voltage and current induced on a secondary coil depends on the ratio of the number of primary windings to the number of secondary windings. This relationship is called the *turns ratio*. If the secondary coil has more windings than the primary coil, then the voltage induced across the secondary coil will be greater than the primary voltage. For example, if the transformer has 1,000 primary windings and 2,000 secondary windings, the turns ratio is 1,000:2,000 (1:2). With 10 Vac applied to the primary, the secondary will output roughly

$$\frac{10}{\left(\frac{1}{2}\right)} \text{ or 20 Vac}$$

Such an arrangement is known as a *step-up* transformer. If the situation were reversed where the primary coil had 2,000 windings with 1,000 windings in the secondary, the turns ratio would then be 2,000:1,000 (2:1). If you apply 30 Vac to the primary, the secondary output would be

$$\frac{30}{\left(\frac{2}{1}\right)} \text{ or } 15 \text{ Vac}$$

This transformer is known as a *step-down* transformer.

Current also is stepped in a transformer, but opposite to the proportion of voltage steps. If voltage is stepped down by the factor of a turns ratio, current is stepped up by the same factor. This relationship ensures that power out of a transformer is about equal to the power into the transformer.

Because inductors are energy-storage devices, they should not dissipate any power by themselves. However, the wire resistance in each coil, combined with natural magnetic losses in the core, *does* allow some power to be lost as heat. Heat buildup is the leading cause of inductor failure. Long-term exposure to heat can eventually break down the tough enamel insulating each winding and cause a short circuit. Short circuits lower the overall resistance of the core so it draws even more current. Breakdown accelerates until the coil is destroyed.

Active components

Diodes, transistors, and integrated circuits make up a much broader and more powerful group of *active* components. Such components are referred to as *active* because each part actually *does* something. Active components use circuit energy to accomplish specific, practical functions. The next part of this chapter provides an overview of common active components, and shows you what each family of parts can do.

Diodes

The classical diode is a two-terminal semiconductor device that allows current to flow in one direction (*polarity*) only, but not in the other. This one-way property is known as *rectification*. As demonstrated in chapter 6, rectification is absolutely essential to the basic operation of every power supply.

Diodes are available in a wide array of case styles as shown in Fig. 2-23. The size and materials used in a diode case will depend on the amount of current that must be carried. Glass-cased diodes, normally made from silicon, are generally used for low-power (or *small signal*) applications. Plastic or ceramic cased diodes are typically used for low or medium power applications like power supplies, circuit isolation, or inductive flyback protection. Diodes also are available in surface-mount packages. A diode has two terminals. The *anode* is its positive terminal, and the *cathode* is its negative terminal. Note that a diode cathode is always marked with a solid stripe or bar.

Whenever you work with rectifier diodes (regardless of the case size or style), you should be concerned with two major diode specifications: forward current (I_f) and peak inverse voltage (PIV). Choose replacement diodes with I_f and PIV values

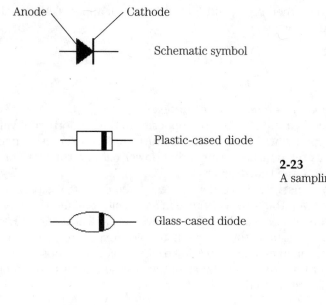

Anode   Cathode

Schematic symbol

Plastic-cased diode

2-23
A sampling of diode case styles.

Glass-cased diode

Surface-mount diode

that closely match the part to be replaced. When a silicon diode is forward biased as shown in Fig. 2-24, the diode develops a constant voltage drop of about 0.6 Vdc. The remainder of applied voltage will drop across the current-limiting resistor. Because a diode dissipates power, it is important that you choose a current-limiting resistor that will restrict forward current to a safe level. If you do not restrict current to a safe level, the diode can be destroyed by excess heat.

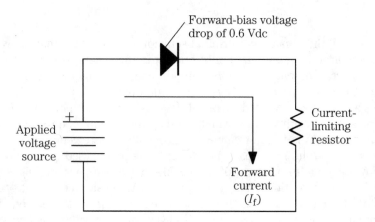

Forward-bias voltage
drop of 0.6 Vdc

Current-
limiting
resistor

Applied
voltage
source

Forward
current
(I_f)

2-24 A forward-biased diode circuit.

A reverse-biased diode, such as the one shown in Fig. 2-25, acts much like an open switch—no current is allowed to flow in the circuit. This characteristic also demonstrates the essential principle of rectification; diode current will flow in only

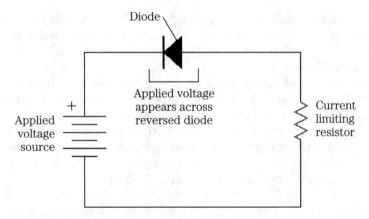

2-25 A reverse-biased diode circuit.

one direction. Whatever voltage is applied across the diode will appear across it. Even if the reverse voltage level were increased, the diode would not conduct. However, there are limits to the amount of reverse voltage that a diode can take. The limit is called *PIV*. If reverse-voltage exceeds PIV, the diode junction can rupture and fail as either an open or short circuit. Typical PIV ratings can easily exceed 200 V.

Although rectifier diodes are not meant to be operated in the reverse-biased condition, the zener diode is a special species designed exclusively for reverse biasing. Figure 2-26 shows a common zener diode circuit. Notice the unique schematic symbol used for zener diodes. When applied voltage is below the zener *breakdown* voltage (typical zener diodes operate at 5, 6, 9, 12, 15, or 24 Vdc), voltage across the zener diode will equal the applied voltage, and no current will flow in the diode. As applied voltage exceeds the zener breakdown voltage, current begins to flow through the diode and voltage across the zener remains clamped at the zener level (that is, 5, 6, 9, 12 Vdc, etc.). Any additional applied voltage is then dropped across

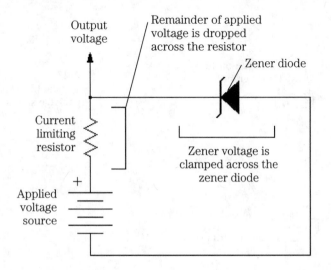

2-26 A basic zener diode regulator circuit.

the current-limiting resistor. As long as applied voltage exceeds the zener break-down voltage, zener voltage will remain constant. This zener action makes zener diodes perfect as simple regulators, and this action is the basis for most methods of linear voltage regulation.

Keep in mind that it is impossible to differentiate between rectifier and zener diodes by their outward appearance—both types of diodes appear identical in every way. The only way you can tell the two types of diodes apart is to look up the particular device in a cross-reference manual, look at the representation of the component in a schematic, or look at the device silk screening on its PC board. Rectifier-type diodes are typically labeled with a *D* prefix (that is, D32, D27, D3, etc.), but zener diodes often use *Z* or *ZD* prefixes (that is, ZD5, ZD201, etc.).

Similarly, it is impossible to discern a faulty diode simply by looking at it, unless the diode has been destroyed by some sudden, severe overload. Such overloads are virtually nonexistent in peripherals such as printers, so you must use test instruments to confirm diode condition. Test instruments are discussed in the next chapter.

In all semiconductor devices, electrons must bridge a semiconductor junction during operation. By modifying the construction of a junction and encapsulating it inside a diffuse plastic housing, electrons moving across a junction will liberate pho-tons of visible (or infrared) light. This action is the basic principle behind light-emit-ting diodes (LEDs). An LED is shown in Fig. 2-27. Notice that an LED is little more than a diode—the wavy arrows indicate that light is moving away from the device. Altering the chemical composition of LED materials will alter the wavelength of

Cathode

Schematic symbol

Large-outline diagram

Cathode mark

Small-outline diagram

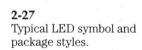

2-27
Typical LED symbol and package styles.

emitted light (i.e. yellow, orange, red, green, blue, infrared, etc.). Like ordinary diodes, LEDs are intended to be forward biased, but LED voltage drops are higher (0.8 to 3.5 Vdc), and LEDs often require 10 to 35 mA (milliamperes) of current to generate the optimum amount of light.

There are two other diode-based devices that you should be familiar with: the SCR (silicon-controlled rectifier) and the triac. You might encounter either of these elements in your printer power supply. An SCR is shown in Fig. 2-28. Notice that SCRs are three-terminal devices. In addition to an *anode* and a *cathode*, a *gate* terminal is added to control the SCR. An ordinary diode turns on whenever it is forward biased. An SCR also must be forward biased, as well as *triggered* by applying a positive voltage (or *trigger signal*) to the gate. Once triggered, an SCR will continue to conduct as long as current is flowing through the SCR. Removing the gate voltage will not stop the SCR from conducting once it has started. If current stops after the gate voltage is removed, the SCR will have to be retriggered. Because SCRs are three-terminal devices, they are often packaged in a fashion very similar to transistors. Most low or medium power SCRs are even packaged in surface-mount packages.

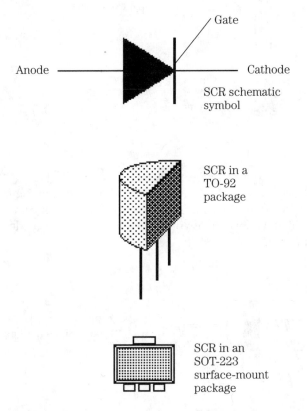

2-28 SCR schematic symbol and package styles.

A triac behaves very much like an SCR, but a triac can be triggered to conduct current in *either* direction through the device instead of only one direction as in the SCR. Figure 2-29 shows a typical triac. Notice that a triac is given TWO anodes because it can conduct in both directions. A triac will conduct once triggered by a voltage applied to its gate lead. The trigger voltage must be the same polarity as the voltage across the triac. For example, if there is a positive voltage from A1 to A2, the trigger voltage must be positive. If there is a negative voltage from A1 to A2, the trigger voltage must be negative. Once triggered, a triac conducts until current stops flowing. After current stops, the triac must be triggered again before it will conduct again.

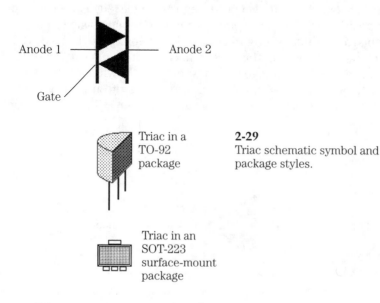

Triac in a TO-92 package

Triac in an SOT-223 surface-mount package

2-29
Triac schematic symbol and package styles.

Diode markings

All diodes carry two very important markings as shown in Fig. 2-30: the part number and the cathode marking. The cathode marking indicates the cathode (or negative) diode lead. Because diodes are polarized devices, it is critical that you know which lead is which. Otherwise, if you incorrectly replace a diode, you might cause a circuit malfunction.

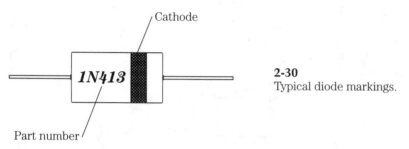

Cathode

1N413

Part number

2-30
Typical diode markings.

Unlike passive components, diode part numbers contain no tangible information on a diode performance specifications or limits. Instead, the part number is an index or reference number that allows you to look up the particular specification in a manufacturer's or cross-reference data book. Classical diode part numbers begin with the prefix *1N*, followed by anywhere from one to four digits. The *1N* prefix is used by the JEDEC (Joint Electron Devices Engineering Council) in the United States to indicate devices with one semiconductor junction (diodes). Classical Japanese diode part numbers begin with the prefix *1SS*, where *SS* means *small signal*.

You will almost certainly encounter diodes with many unique and arcane markings. Fortunately, there are many clues to guide you along. Your first clue will be the white (or gray) cathode band—at least you can identify the part as a diode. The second identifier will be the silk-screen lettering on the PC board. Diodes are usually assigned *D* or *ZD* numbering prefixes to denote a rectifier-type or zener diode. Once you are confident that you have identified a diode, use a cross-reference index to look up the suspect part replacement or equivalent. The specifications you find for the equivalent part in a cross-reference manual will closely (if not exactly) match those for the original part. There are many semiconductor cross-references available.

Transistors

A *transistor* is a three-terminal semiconductor device whose output signal is directly controlled by its input signal. With passive components, a transistor can be configured to perform either amplification or switching tasks. Unfortunately, there is just not enough room in this book to discuss the theory and characteristics of transistors, but it is important that you know the most important concepts of transistors, and understand their various uses.

There are two major transistor families: bipolar and FET (field-effect transistor). Bipolar transistors are common, inexpensive, general-purpose devices that can be made to handle amplification and switching tasks with equal ease. The three leads of a bipolar transistor are the base, emitter, and collector. In most applications, the *base* serves as the transistor input—that is where the input signal is applied. The *emitter* is typically tied to ground (usually through one or more values of resistance), and the *collector* provides the output signal. The transistor also can be configured with the input signal supplied to the collector, the base is grounded, and the output appears on the emitter.

There are two species of bipolar transistor: NPN and PNP. For an NPN transistor, base and collector voltages must be positive (with respect to the emitter). As base voltage increases, the transistor is turned on and current begins to flow from collector to emitter. As base voltage increases further, the transistor continues to turn on and allow more current into the collector until the transistor finally saturates. A saturated transistor cannot be turned on any further. A PNP transistor requires negative base and collector voltages (with respect to the emitter) to cause the transistor to turn on and conduct current. As base voltage becomes more negative, the transistor turns on harder until it saturates. By far, NPN transistors are more commonly used in small electronics.

When a bipolar transistor is used in a switching circuit (also known as a *driver* circuit), the arrangement is often simpler as shown in Fig. 2-31. Unlike a transistor amplifier whose output signal varies in direct proportion to its input, a switching circuit is either on or off—very much like a mechanical switch. Laser printers make extensive use of switching circuits to operate LEDs, solenoid clutches, and motors. The

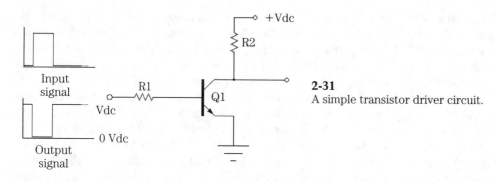

Input
signal

Output
signal

2-31
A simple transistor driver circuit.

typical computer switching circuit uses a digital signal from an IC as the control signal. When the control signal is at a logic 0, the transistor (and its load) remains off. When a logic 1 signal is supplied to the driver, the transistor turns on fully (saturates). Current flows through the load, into the collector, through the transistor to the emitter, then to ground. A base resistor is added to limit current into the base from the signal source. An additional resistor might need to be placed in the collector circuit to limit current if the load is too large.

Phototransistors are a unique variation of bipolar transistors. Instead of an electrical signal being used to control the transistor, photons of light provide the base signal. Light enters the phototransistor through a clear quartz or plastic window on the transistor body. Light that strikes the transistor base liberates electrons that become base current. The more light that enters the phototransistor, the more base current is produced, and vice versa. Although phototransistors can be operated as linear amplifiers, they are most often found in switching circuits that detect the absence or presence of light.

Although phototransistors can detect light from a wide variety of sources, it is normal to use an LED to supply a known, constant light source. When a phototransistor and LED are matched together in this way, an *optocoupler* (or *optoisolator*) is formed as shown in Fig. 2-32. Notice the new schematic symbol for a phototransistor. The wavy lines indicate that light is entering at the base. Optocouplers are used in printers to detect the presence or absence of paper, or to detect whether the printer enclosures are closed completely. Optocouplers also can be fabricated together on the same integrated circuit to provide circuit isolation.

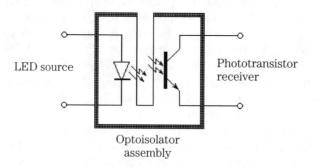

LED source

Phototransistor
receiver

Optoisolator
assembly

2-32 Cross-sectional diagram of an optoisolator.

Field-effect transistors (FETs) are constructed in a radically different fashion from bipolar transistors. Although FETs make use of the same basic materials and can operate as either amplifiers or switches, they require biasing components of much higher value to set the proper operating conditions. FETs are either N-channel or P-channel devices as shown in Fig. 2-33. The difference in transistor types depends on the materials used to construct the particular FET. An FET has three terminals: a *source*, a *gate*, and a *drain*. These leads correspond to the emitter, base, and collector of a bipolar transistor. The gate is typically used for the input or control signal. The source is normally tied to ground (sometimes through one or more values of resistance), and the drain supplies the output signal.

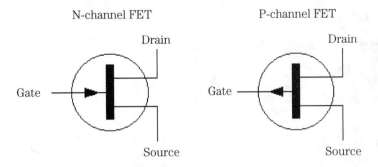

N-channel FET   P-channel FET

2-33 FET Schematic symbols.

When no voltage is applied to an FET gate, current flows freely from drain to source. Any necessary current limiting must be provided by inserting a resistor in series into the drain or source circuit. By adjusting the control voltage at the gate, current flow in the drain and source can be controlled. For an N-channel FET, control voltage must be a negative voltage. As gate voltage is made more negative, channel current is restricted further until it is cut off entirely. For a P-channel FET, a positive control voltage is needed. Higher positive gate voltage restricts channel current further until the channel is cut off.

A variation of the FET is the MOSFET (metal-oxide semiconductor FET). It is unlikely that you will ever encounter discrete MOSFET devices in your small computer, but many sophisticated digital integrated circuits make extensive use of MOSFETs. One of the few undesirable characteristics of FETs and MOSFETs is their sensitivity to damage by electrostatic discharge (ESD). You can learn about ESD in chapter 4.

There are a variety of electrical specifications that describe the performance and characteristics of particular transistors. When you discover that a transistor must be replaced, it is always wise to use an *exact* replacement part. That way, you are sure the replacement part will behave as expected. Under some circumstances, however, an exact replacement part might take too long to get, or not even be available. You can then use manufacturer's or cross-reference data to locate substitute parts with specifications similar to the original part. Keep in mind that substituting parts with a different part—even when specifications are very similar—might have an unforeseeable impact on the circuit. Do not attempt to use "close" replacement parts unless you have a keen understanding of transistor principles and specifications, and you know exactly what the part does in a circuit.

Transistors are available in a wide variety of case styles and sizes depending on the power that must be handled. Figure 2-34 shows a selection of five popular case styles. Low-power, general-purpose devices are often packaged in the small, plastic TO-92 cases. The TO-18 metal case also is used for low-power devices, but the TO-18 shown houses a phototransistor. Note the quartz window on the case top that allows light to enter the device. For regular transistor applications, the TO-18 "top hat" case is all metal.

TO-18 case
(low power)
phototransistor
opening shown

TO-92 case
(low power)

TO-220 case
(medium to
high power)

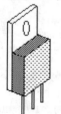

SOT-223
surface-mount
package

2-34
Typical transistor case styles.

SOT-143
surface-mount
package

Medium-power transistors use the larger plastic TO-128 or TO-220 cases. The TO-128 uses a thin metal heat sink molded into the top of the device. TO-220 cases use a large metal mounting flange/heat sink located directly behind the plastic case. The flange provides mechanical strength, as well as a secure thermal path for an external heat sink. An all-metal TO-3 case (not shown) is used for high-power transistors. Two mounting holes are provided to bolt the device to a chassis or external heat sink. Usually, case size is proportional to the power capacity of the transistor.

Transistors also can be manufactured in surface-mount cases. Two typical surface-mount small-outline transistor (SOT) case styles are shown in Fig. 2-34. Due to their small size, SMT transistors cannot dissipate very much power, but they are ideal for small, low-power systems such as computers or printers.

Transistor markings

As shown in Fig. 2-35, a transistor part number is merely an index or reference number that allows you to look up the equivalent components or specifications for a part in a data book or cross-reference manual. The number itself contains no useful information about the actual performance characteristics or limits of the part. Classical bipolar transistor part numbers begin with the prefix *2N*, followed by up to five digits. The *2N* prefix is used by JEDEC in the United States to denote devices with two semiconductor junctions. Classical Japanese transistor part numbers begin with any of four prefixes: *2SA* (high-frequency PNP transistor), *2SB* (low-frequency PNP transistor), *2SC* (high-frequency NPN transistor), and *2SD* (low-frequency NPN transistor). JEDEC uses the prefix *3N* to denote FETs or Junction FETs (JFETs). The prefixes *2SJ* (P-channel JFET), *2SK* (N-channel JFET), and *3SK* (N or P channel MOSFETs) have been used in Japan.

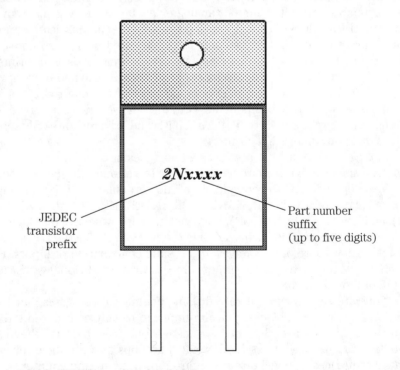

2-35 JEDEC transistor markings.

You also might encounter many transistors with arcane or nonstandard markings. In almost all cases, you can identify replacement transistors and look up performance specifications using manufacturer's data or a cross-reference guide. Although the specifications found in a cross-reference guide are for the replacement parts, they will generally match the original part specifications very closely.

As with diodes, transistors rarely show any outward signs of failure unless they have melted or shattered from an extreme overload. You generally must use test equipment to identify faulty transistors. Testing can be accomplished by measuring the device while the circuit is running, or removing the device from the circuit and measuring its characteristics out of circuit.

Integrated circuits

Integrated circuits (ICs) are by far the most diverse and powerful group of electronic components that you will ever deal with. They have rapidly become the fundamental building blocks of modern electronic circuits. Amplifiers, memories, microprocessors, digital logic arrays, oscillators, timers, regulators, and a myriad of other complex functions can all be manufactured as ICs. Circuits that only a decade ago would have required an entire PC board in a printer are now being fabricated on a single IC. Although you can often estimate the complexity (and importance) of an integrated circuit from the number of pins that it has, it is virtually impossible to predict precisely what an IC does just by looking at it. You will need a schematic of the circuit or manufacturer's data of a particular IC to determine what the IC does.

Every integrated circuit—whether analog or digital—usually is made up of microscopic transistors, diodes, capacitors, and resistors that are fabricated onto an IC die. Many capacitors and inductors cannot be fabricated on ICs, so conventional parts can be attached to an IC through one or more external leads. Your printer is almost entirely a digital system. That is, most of the ICs are designed to work with binary signals. The microprocessor, memory, and most of the controller ICs in your laser printer, are digital logic components. Other ICs, however, are intended to work with analog signal levels. Serial communication ICs and driver ICs are only some analog devices that you will find in a printer.

A logic *gate* is a circuit that produces a binary result based on one or more binary inputs. A single integrated circuit can hold as few as one logic gate or thousands of logic gates depending on the complexity of the particular IC. There are eight basic types of logic gates; AND, OR, NAND, NOR, INVERTER, BUFFER, XOR (exclusive OR), and XNOR (exclusive NOR). Each gate uses its own particular logic symbol as shown in Fig. 2-36. Beside each symbol is the *truth table* for the particular gate. A truth table shows gate output for every possible combination of inputs. For the sake of simplicity, no more than two inputs are shown, but individual gates can have four, eight, or more inputs.

Flip-flops are slightly more involved than general-purpose gates, but they are such flexible logic building blocks, that flip-flops are usually considered to be logic gates. In the simplest sense, flip-flops are memory devices, because they can "remember" the various logic states in a circuit. Flip-flops also are ideal for working with logical sequences. You will often find flip-flops used around counter-timer circuits. There are three classical variations of flip-flops: the D flip-flop, the SR flip-flop, and the JK flip-flop.

As you look over the symbols in Fig. 2-36, you might notice that some of the inputs have a *bubble* at the device. Whenever you see a bubble, it indicates active-low logic. In conventional (active-high) Boolean logic, a *true* signal indicates an ON condition, or the presence of a voltage. In active-low logic, a true signal is OFF, and volt-

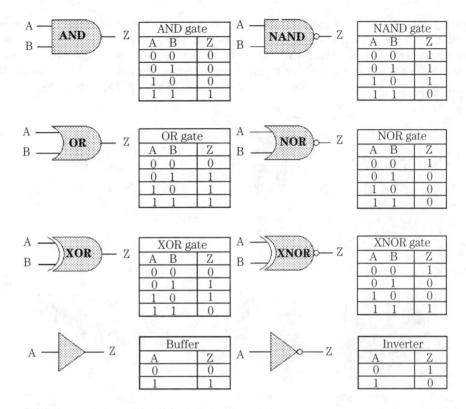

2-36 A comparison of basic logic gates.

age is absent. This concept will be extremely important in troubleshooting so that you do not confuse a correct signal for an erroneous signal.

The vast majority of logic components used in today's small computers contain so very many gates that it would simply be impossible to show them all on a schematic diagram. Current microprocessors, gate arrays, and application-specific ICs (ASICs) can each contain thousands of gates. To simplify schematics and drawings, most highly integrated ICs are shown only as generic logic blocks that are interconnected to one another.

Every logic IC requires a power source to operate. At least one positive voltage source V_{cc} must be applied to the IC. The IC also must be grounded with respect to the source voltage. An IC's ground pin is usually labeled Vss or GND. Since the days of the first logic ICs, supply voltage has been a standard of +5 Vdc. Using a +5 Vdc source, logic 1 outputs are interpreted as +2.4 Vdc or higher, and logic 0 outputs are considered to be +0.4 Vdc or lower. The transistor circuitry within each logic gate accepts inputs at these levels. If +5 Vdc is not supplied to the IC, it will function erratically (if at all). If more than +6 Vdc is forced into the IC, excess power dissipation will destroy the IC.

Integrated circuits are manufactured in a staggering array of package styles. Older package styles such as the *DIP* (dual in-line package) and *SIP* (single in-line package) are intended for printed circuits using conventional (through-hole) assembly techniques. Through-hole assembly requires that a component be inserted so that its leads protrude through the PC board. The leads are soldered into place where each lead protrudes. However, the tendency to pack more powerful ICs onto smaller PC boards has given rise to an overwhelming number of surface-mount IC package styles. Figure 2-37 shows a small sampling of typical package styles.

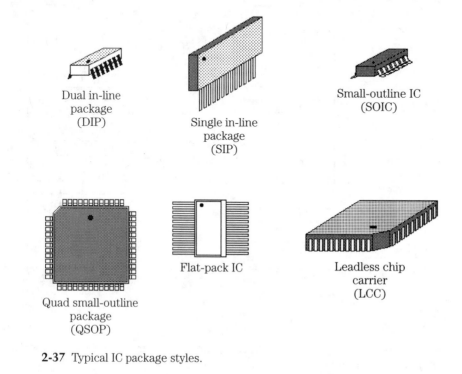

Dual in-line package (DIP)

Single in-line package (SIP)

Small-outline IC (SOIC)

Quad small-outline package (QSOP)

Flat-pack IC

Leadless chip carrier (LCC)

2-37 Typical IC package styles.

One of the key advantages of surface-mount packages is that components can be soldered onto *both* sides of a PC board—thus, almost doubling the amount of available PC board area. The *SOIC* (small-outline IC) design appears very similar to a DIP, but the SOIC is significantly smaller, and its flat (or gull-wing shaped) pins are spaced together very closely. The *VSOP* (very small-outline package) and the *QSOP* (quad small-outline package) are designed to package complex ICs into extremely small packages with leads on two or four sides. Quad packages also are square (instead of rectangular) with pins on all four sides. *SOJ* (small-outline J-lead) packages replace regular gull-wing leads with leads that are bent down and under the device in a *J* shape.

Flat packs tend to be large, square ICs used primarily for more sophisticated functions such as microprocessors and specialized controllers. A *QFP* (quad flat pack) and a *TQFP* (thin quad flat pack) offer as many as 100 pins (25 pins per side). TQFPs are handy in extremely tight spaces where regular QFPs might get in the way of other assemblies. Chip carriers are either leaded (with leads) or unleaded (without leads). *LLC* (leadless chip carriers) simply provide exposed contacts that require a chip-carrier socket to guarantee proper connections. *PLCC* (plastic leaded chip carrier) packages offer J-shaped pins that can either be surface mounted directly to a PC board or inserted into a chip-carrier socket. The *PGA* (pin grid array) is one of the most sophisticated packaging schemes in use today. PGAs can provide hundreds of leads on an IC. Sophisticated microprocessors such as Intel's 80486 are packaged in PGAs exceeding 150 pins. PGAs also require the use of sockets to ensure proper contact for all pins. **Take extreme care to prevent damage to pins when inserting and extracting PGAs.**

It is exceptionally rare for any type of IC to show outward signs of failure, so it is very important for you to check carefully any suspect ICs using appropriate test equipment and data while the IC is actually operating in the system. Gather all the information you can about an IC before replacing it, because IC replacement always carries an element of risk. You risk damage to the PC board during IC removal, and damage to the new IC during installation.

3
CHAPTER

Soldering and test instruments

Up to now, you have seen the features and specifications of laser printers, with a variety of typical subassemblies. You know about their major electrical and mechanical components (Fig. 3-1). This chapter introduces you to the tools and test equipment needed for successful laser printer troubleshooting. If you are unfamiliar with test equipment, take some time and read this chapter carefully. Experienced troubleshooters can skip this chapter, but might need to refer to it later for reference.

Tandy Corporation.

3-1 A Tandy LP400 laser printer.

Small tools and materials

It might sound strange, but hand tools can often make or break your repair effort. If you have ever started a repair and been delayed by one missing screwdriver, pliers, or wrench (you knew it was there last week), then you know how much time and frustration can be saved by gathering the proper tools *before* beginning—not after. Check your toolbox now! If you do not have a toolbox just yet, now is a good time to consider the supplies and tools you need for mechanical and electronic repairs.

Hand tools

Screwdrivers are always good tools to begin with. Figure 3-2 shows five types of screw heads. A healthy variety of both regular and Phillips-type screwdrivers will go a long way for basic assembly, disassembly, and adjustment tasks. Avoid excessively large or unusually small screwdrivers unless there is a specific need for them. Several short-shaft screwdrivers can come in handy when working in confined areas—printer assemblies can be very densely packaged. Allen-head (hexagonal hole) screws also are common, so include a set of Allen keys in your toolbox. Watch for specialty screws. Spline and Torx-type screws are growing in popularity as manufacturers seek to keep untrained personnel out of their equipment. Large mail-order and retail hardware stores usually stock Spline and Torx drivers.

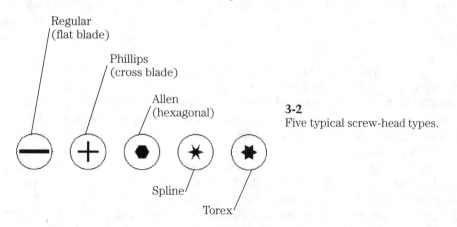

3-2
Five typical screw-head types.

Consider three types of pliers. A garden variety pair of mechanic's pliers is useful for keeping tight hold on any nut, bolt, or other pesky part. Two pairs of needle-nose pliers (one short nose, one long nose) will round out the collection. Needle-nose pliers are great for grabbing and holding parts in tight spaces.

A set of small, electronics-grade, open-end wrenches work well to hold small nuts during assembly and disassembly. Wrenches should have thin bodies (for tight spaces) and should be below ⁵⁄₁₆ inch (or 8 mm for metric sizes). If you have the choice between metric or U.S. wrenches, get metric. Most laser printers are made in the Pacific Rim (that is, Japan, Korea, Singapore, Hong Kong, and so on) where metric parts are standard. A small adjustable wrench is always a good addition to a toolbox.

Because you are working with electronic systems, have on hand two sizes of wire cutters (one small and one medium) to cut jumper wires or trim replacement component leads. Include a separate wire stripper to remove insulation from wires or components. Resist the temptation to strip insulation using wire cutters. Even if insulation is removed successfully, cutting blades can often leave a nick or pinch in the conductor that might later fatigue and break. If you plan to make or repair connectors, you might need special crimp or insertion tools to do the job. Use your judgment regarding exactly what tools you need for connector repair.

Any wiring or circuit work is going to require a *soldering iron*. Invest in a good-quality soldering *pencil*—in the range of 20 to 25 W (30 W maximum). Avoid heavy-duty, high-power soldering *guns*. They might work just fine for plumbers, but that much power can easily destroy printed circuit boards, wire insulation, and components. Always be sure to have one or two spare soldering tips on hand.

Remember that soldering irons exceed 500°F at the tip. Not only are these temperatures a serious hazard to your personal safety, but they are a serious fire hazard as well. Be sure to park a hot soldering iron in a strong, wire-frame holder. NEVER leave an iron unattended on a bench or table.

Materials

To use a soldering iron, you also are going to need an ample supply of solder. Use a 60% tin, 40% lead (60/40) solder containing a rosin cleaning agent. Never use paste flux containing acids or solvents, or use solder containing acid flux (sometimes called *acid core or plumber's solder*). Harsh solvents destroy delicate component leads and circuit traces.

A spool of *hook-up wire* can serve a variety of uses ranging from printed circuit repair and wire splices, to make shift test leads. Solid wire (18 to 20 gauge) is often easiest to work with, but stranded wire can be used just as well. You might find it easier to keep several smaller spools of different-colored wire so you have color-coded wire for different purposes (for example, red for dc voltage, blue for dc ground, and yellow for signal wires).

Include a set of *alligator leads*. Alligator leads are available in a selection of lengths, colors, and wire gauges. Alligator leads are handy for temporary jumping of signals during a repair, or to test the reaction of another signal in a circuit. A set of small alligator leads and a set of large alligator leads will cover you under most circumstances. Individual *alligator clips* make excellent heat sinks to conduct heat away from components when you solder or desolder.

Heat-shrink tubing provides quick, clean, and effective insulation for exposed wiring or splices. Most general-purpose electronics supply stores sell heat shrink tubing in 3 foot lengths, or in packages of assorted sizes. Use a heat gun to shrink the tubing, although you could use a blow drier. Heat-shrink tubing is available in a selection of colors.

These tools and materials are only a few more common supplies that you will need to get started. The list is by no means complete. Hundreds of general and special purpose tools are available to help you in your repairs—far too many to cover completely. Experience is your best guide in deciding which tools and materials are best for you.

Soldering

Soldering is the most commonly used method of connecting wires and components in an electrical or electronic circuit. In soldering, metal surfaces (component leads, wires, or printed circuit traces) are heated to high temperatures, then joined with a layer of compatible molten metal. When done correctly, soldering forms a lasting, corrosion-resistant, intermolecular bond that is mechanically strong and electrically sound. All that is needed is an appropriate soldering iron and electronics-grade (60/40) solder. This section covers both regular soldering and surface-mount soldering.

Soldering background

By strict definition, *soldering* is a process of bonding metals together. There are three distinct types of soldering: brazing, silver soldering, and soft soldering. Brazing and silver soldering are used for hard or precious metals, but soft soldering is the technique of choice for electronics work.

To bond wire and component leads (typically made of copper), a third metal must be added while in its molten state. The bonding metal is known simply as *solder*. Several different types of solder are available to handle each soldering technique, but the chosen solder must be compatible with the molecular characteristics of the metals to be bonded—otherwise a bond will not form. Lead and tin are two common, inexpensive metals that adhere very well to copper. Unfortunately, neither metal by itself has the strength, hardness, and melting-point characteristics to make them useful. Therefore, lead and tin are combined into an alloy. A ratio of approximately 60% tin and 40% lead yields an alloy that is pliable, has reasonable hardness, and a low melting point that is ideal for electronics work. This mixture is the solder that you should use.

Although solder adheres very well to copper, it does not adhere well at all to the oxides that form naturally on copper. Although conductors might *look* clean and clear, some amount of oxidization is always present. Remove oxides before soldering to get a good bond. Apply a rosin cleaning agent (called *flux*) to conductors before soldering. Although rosin is chemically inactive at room temperature, it becomes extremely active when heated to soldering temperatures. Active flux combines with oxide and strips it away, leaving clean copper surfaces for molten solder. As a completed solder joint cools, any residual rosin cools and returns to an inactive state. Never use acid or solvent-based flux to prepare conductors. They can clean away oxides as well as rosin, but acids and solvents remain active after the joint cools. Over time, the residue will dissolve copper wires and connections, and eventually cause a circuit failure. You can get rosin flux as a paste that can be brushed into conductors before soldering, but most electronic solders have a core of rosin manufactured right into the solder strand itself. The rosin cleans the joint as solder is applied. Such *rosin-core* solder is much more convenient than working with flux paste.

Irons and tips

A soldering iron is little more than a resistive heating element built into the end of a long steel tube as shown in the cross-sectional diagram of Fig. 3-3. When ac volt-

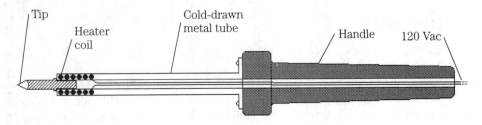

3-3 Cross-sectional view of a typical soldering iron.

age is applied to the heater, it warms the base of a tip. Any heat conducted down the cool-down tube is dissipated harmlessly to the surrounding air. Dissipating the heat keeps the handle temperature low enough to hold comfortably.

Most of the heat is channeled into a soldering tip similar to the one shown in Fig. 3-4. Tips often have a core of copper that is plated with iron. The tip is coated with a layer of nickel to stop high-temperature corrosion, then plated with chromium. A chromium coating renders the tip *nonwettable*—solder will not stick. Because solder must stick at the tip end, that end is plated with tin. A tin coating makes the end *wettable*. Tips are available in a wide range of shapes and sizes. Before you select the best tip for the job, you must understand ideal soldering conditions.

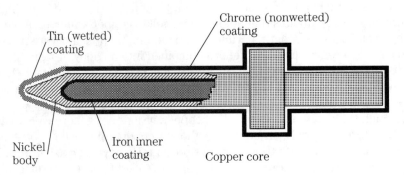

3-4 Soldering iron tip.

The very best electronic soldering connections are made within only a narrow span of time and temperature. A solder joint heated between 500 to 550°F for 1 to 2 seconds will make the best connections. You must select your soldering iron wattage and tip to achieve these conditions. The entire purpose of soldering irons is not to melt solder. Instead, a soldering iron is supposed to deliver heat to a joint—the *joint* will melt solder. A larger joint (with more numerous or larger conductors) requires a larger iron and tip than a small joint (with fewer or smaller conductors). If you use a small iron to heat a large joint, the joint can dissipate heat faster than the iron can deliver it, so the joint might not reach an acceptable soldering temperature. Conversely, using a large iron to heat a small joint will overheat the joint. Overheating can melt wire insulation and damage printed circuit board runs. Match wattage to the application. Most general-purpose electronics work can be soldered using an iron with a rating below 30 W.

Because the end of a tip actually contacts the joint to be soldered, its shape and size can assist heat transfer. When heat must be applied across a wide area (such as a wire splice), a wide area tip should be used. A screwdriver (or *flat blade*) tip such as shown in Fig. 3-5 is a good choice. If heat must be directed with pinpoint accuracy for small, tight joints or printed circuits, a narrow blade or conical tip is the best. Two tips for surface-mount desoldering also are shown in Fig. 3-5.

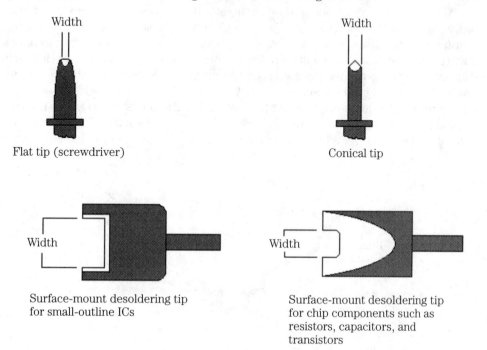

Flat tip (screwdriver)

Conical tip

Surface-mount desoldering tip
for small-outline ICs

Surface-mount desoldering tip
for chip components such as
resistors, capacitors, and
transistors

3-5 An assortment of conventional and surface-mount tips.

Soldering

SAFETY FIRST—always park your soldering iron in a secure holder while it is on. Never allow a hot iron to sit freely on a table or anything flammable. Make it a point to ALWAYS wear safety glasses when soldering. Active rosin or molten solder can easily flick off the iron or joint and do permanent damage to the tissue in your eyes.

Give your soldering iron plenty of time to warm up (5 minutes is usually adequate). Once the iron is at its working temperature, you should coat the wettable portion of the tip with fresh solder (this is known as *tinning* the iron). Rub the tip into a sponge soaked in clean water to wipe away any accumulations of debris or carbon that might have formed, then apply a thin coating of fresh solder to the tip end. Solder penetrates the tip to a molecular level and forms a cushion of molten solder that aids heat transfer. Re-tin the iron any time its tip becomes blackened—perhaps every few minutes or after several joints.

Tin each conductor before actually making the complete joint. To tin a wire, prepare it by stripping away $3/16$ to $1/4$ inch of insulation. As you strip insulation, be sure not to nick or damage the conductor. Heat the exposed copper for about 1 second, then apply solder into the wire—not into the iron. If the iron and tip are appropriate, solder should flow evenly and smoothly into the conductor. Apply enough solder to bond each exposed strand of a stranded wire. When tinning a solid wire or component lead, apply just enough solder to lightly coat the conductor surface. Conductors heat faster and solder flows better when all parts of a joint are tinned in advance.

Making a complete solder joint is just as easy. Bring together each of your conductors as necessary to form the joint. For example, if you are soldering a component into a printed circuit board, insert the component leads into their appropriate locations. Place the iron against all conductors to be heated as shown in Fig. 3-6. For a printed circuit board, heat the printed trace and component lead together. After about 1 second, flow solder gently into the conductors—not the iron. Be sure that solder flows cleanly and evenly into the joint. Apply solder for another 1 or 2 seconds; then remove both solder and iron. Do not touch or move the joint until molten solder has set after several seconds. If the joint requires additional solder, reheat the joint and flow in a bit more solder.

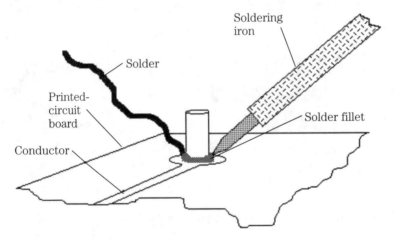

3-6 Soldering a typical printed-circuit junction.

You can identify a good solder joint by its smooth, even, silvery-gray appearance. Any charred or carbonized flux on the joint indicates that your soldering temperature is too high (or that heat is being applied for too long). Remember that solder can not flow unless the joint is hot. If it is not, solder will cool before it bonds. The result is a rough, built-up, dull-gray or black mound that does not adhere very well. The mound is known as a *cold* solder joint. You can usually correct a cold joint by reheating the joint properly and applying fresh solder.

Surface-mount soldering

Conventional printed-circuit boards use through-hole components. Parts are inserted on one side of the PC board, and their leads are soldered to printed traces on the other side. Surface-mounted components do not penetrate a PC board. Instead, the components rest on only one side of the board as shown in Fig. 3-7. Metal tabs replace component lead wires. Surface-mount components range from discrete components such as resistors and capacitors to active parts such as transistors and inte-

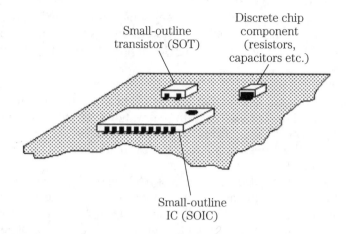

Small-outline
transistor (SOT)

Discrete chip
component
(resistors,
capacitors etc.)

Small-outline
IC (SOIC)

3-7 Simplified view of a surface-mount PC board.

grated circuits. Even sophisticated ICs like microprocessors and ASICs can be found in surface-mount packages.

During manufacture, surface-mount parts are glued into place on a PC board, then the board is brought quickly up to soldering temperature in a special chamber. Molten solder is flowed over the board where it adheres to heated component leads and PC traces. The remainder of the board is chemically and physically masked before soldering to prevent molten solder from sticking elsewhere. The finished board is then cooled slowly to prevent thermal shock to the components, masks are stripped away, and the board can be tested (or any through-hole parts can be added). This type of fabrication is called *wave soldering*, and it is similar to the principle used to mass-solder through-hole PC boards. A different manufacturing technique applies a layer of solder paste to a masked PC board before components are applied, then the board is heated to flow solder into PC trace. After components are glued into place, the board is quickly reheated so solder will adhere to each component lead. The finished board is then cooled slowly. This process is known as *reflow soldering*. Figure 3-8 shows a close-up view of an SMT (surface-mount technology) solder connection.

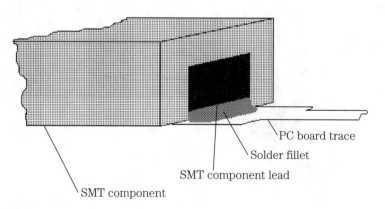

PC board trace

Solder fillet

SMT component lead

SMT component

3-8 Close-up view of an SMT solder connection.

Although the specific methods of surface-mount soldering will have little impact on your troubleshooting, you should understand how surface-mount components are assembled so that you can disassemble them properly during your repair.

Desoldering

Ideally, desoldering a connection involves removing the intermolecular bond that has been formed during soldering. In reality, however, this is virtually impossible. The best that you can hope for is to remove enough solder to gently break the connection apart without destroying the joint. Desoldering is a game of removing as much solder as possible.

You will find some connections very easy to remove. For instance, a wire inserted into a printed circuit board can be removed just by reheating the joint and gently withdrawing the wire from its hole once solder is molten. You can use desoldering tools to clear away the solder itself after the connection is broken.

Surface-mount components present a special problem, because it is impossible to move the part until it is desoldered completely. By using special desoldering tips as shown in Fig. 3-5, you can heat all leads simultaneously so the part can be separated in one quick motion. There also are special tips for desoldering a selection of IC packages. Once a part is clear, excess solder can be removed with conventional desoldering tools such as a *solder vacuum* or *solder wick*.

Desoldering through-hole components is not as easy as it looks. You must heat each solder joint in turn, and use a desoldering tool to remove as much solder as possible. Once each lead is clear, you will probably have to break each lead free as shown in Fig. 3-9. Grab hold of each lead and wiggle it back and forth gently until it breaks free. An alternate method is to heat each joint while withdrawing the lead with a pair of needle-nose pliers, then clean up any excess solder later. Unfortunately, this process cannot be used with all components. Experience will teach you the finer points of desoldering.

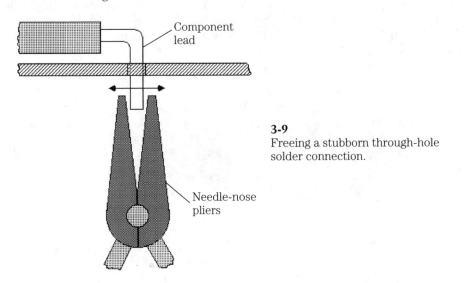

3-9
Freeing a stubborn through-hole solder connection.

Multimeters

Test meters can go by many names. Some people call them *multimeters* or just *meters*, and others might refer to them as *VOMs* (volt-ohm-milliammeter) or *multitesters*. Regardless of what name you choose to call them, multimeters are the handiest and most versatile piece of test equipment that you will ever use. If your toolbox does not contain a good-quality multimeter yet, now is a good time to consider purchasing one.

Even the most basic multimeters can measure ac and dc voltage, current, and resistance. For less than $150, you can buy a good digital multimeter that also includes features like a capacitance checker, continuity checker, diode checker, and transistor checker. Digital multimeters are easier to read, more tolerant of operator error, and more precise than analog multimeters. Figure 3-10 shows a typical digital multimeter.

3-10
A B+K Model 2707 DVM.

Consider two factors when you use a multimeter. First, the meter must be set to the desired *function* (voltage, current, capacitance, etc.). Second, the *range* must be set properly for that function. If you are unsure what range to use, start by choosing the highest possible range. Once you have a better idea of what readings to expect, the range can be reduced to achieve a more precise reading. If your signal exceeds the meter range, an over range warning will be displayed. Many digital multimeters are capable of selecting the proper range automatically (auto-ranging).

You can use a multimeter for two types of testing: static and dynamic. *Dynamic* tests are made with power applied to a circuit, and *static* tests are made on unpowered circuits or components. Measurements like voltage, current, and frequency are dynamic tests; most other tests such as resistance/continuity, capacitance, diode and transistor junction quality are static tests. The following is a review of basic multimeter measurement techniques.

Measuring voltage

Multimeters can measure both dc voltage (marked DCV or Vdc) and ac voltage (marked ACV or Vac). Remember that all voltage (either ac or dc) must be measured *in parallel* with the desired circuit or component. Never interrupt a circuit and attempt to measure voltage in series with other components. Any such reading would be meaningless, and your circuit might not even function.

Set your multimeter to its appropriate function (DCV or ACV), then select the proper range. If you are unsure what range to use, start at the *highest* range to prevent possible damage to the meter. An auto-ranging multimeter will select its own range. Place your test leads across (in parallel) with the part under test as shown in Fig. 3-11, and read voltage directly from the digital display. The dc voltage readings are polarity sensitive, so if you read +5 Vdc and then reverse the test leads, you will see –5 Vdc. The ac voltage readings are not polarity sensitive.

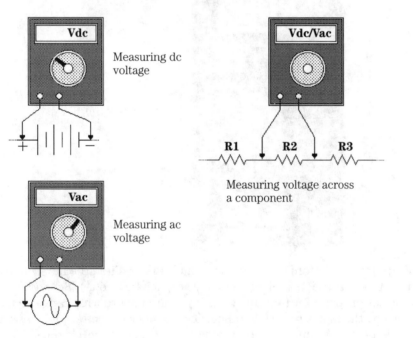

3-11 Measuring voltage.

Measuring current

Most general-purpose multimeters allow you to measure ac current (ACA or Iac) and dc current (DCA or Idc) in a circuit, although there are often few ranges to choose from. As with voltage measurements, current is measured in a working circuit with power applied, but current must be measured *in series* with the circuit or component under test. Inserting a meter in series, however, is not always an easy task. In many cases, you must physically interrupt a circuit at the point you wish to measure, then connect test leads across the break. Although it might be easy to interrupt a circuit, keep in mind that you also must put the circuit back together, so use care when choosing a point to break. **Never** try to read current in parallel. Current meters, by their nature, exhibit very low resistance across their leads (sometimes below 0.1 Ω). Placing a current meter in parallel can cause a short circuit across a component that can damage the part, the circuit under test, or your multimeter.

Set your multimeter to the desired function (DCA or ACA) and select the appropriate range. If you are unsure about proper range, set the meter to its *highest* range. You might need to plug one of your test leads into a different current input jack on the meter. Unless your multimeter is protected by an internal fuse, it can be damaged by excessive current. Make sure that the meter can handle the amount of current you are expecting.

Turn off all power to a circuit before inserting a current meter. This precaution prevents unpredictable circuit operation. If you wish to measure power-supply current feeding a circuit as shown in Fig. 3-12, open the power supply line at any convenient point (often at the supply or circuit board connectors). Insert the meter and reapply power. Read current directly from the display. This procedure also can be used for measuring current within a circuit.

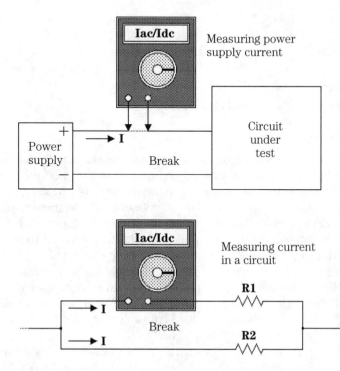

3-12 Measuring current.

Measuring frequency

Some multimeters offer a frequency counter (f or Hz) that can read frequency directly. The ranges available depend on your particular meter. Simple hand-held meters can often read up to 100 kHz, and bench-top multimeters can handle 10 MHz or more. Frequency measurements are dynamic readings made *in parallel* across a component or circuit.

Set your multimeter to its frequency counter function and select the appropriate range. If you are unsure just what frequency to expect, start at its *maximum* frequency range. Place your test leads across the signal source as shown in Fig. 3-13, and read frequency directly from the display.

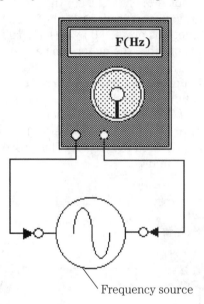

3-13
Measuring frequency.

Measuring resistance

Resistance (in ohms) is the most common static measurement that you can take with a multimeter. Measuring resistance is a handy function, not only for checking resistors, but for checking other resistive elements such as wires, connectors, motors, solenoids, and some semiconductor components. Resistance is measured *in parallel* across components with all circuit power *OFF* as shown in Fig. 3-14. You might need to remove at least one component lead from its circuit to prevent interconnections with other components from causing false readings.

Ordinary resistors can be checked simply by switching to a resistance function and selecting the proper range. Many multimeters can reliably measure resistance up to 20 MΩ. Place your test leads across the component and read resistance directly from the display. If resistance exceeds the selected range, the display will indicate an over range or infinite resistance condition.

Continuity checks are made to ensure a reliable, low-resistance connection between two points. For example, you could check the continuity of a cable between two connectors to ensure that both ends are connected properly. Set your multime-

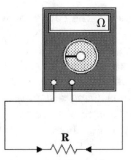

3-14
Measuring resistance.

ter to a low resistance scale, then place your test leads across both points to measure. Ideally, good continuity should be about 0 Ω.

Checking a capacitor

There are two methods of checking a capacitor using your multimeter. If your meter has a built-in capacitance checker, all you need to do is select the capacitance function and set the desired range. You might have to place test probes *in parallel* across the capacitor under test, or you might have to remove the capacitor from the circuit and insert it into a special fixture on the meter face. A capacitance checker will usually display capacitance directly in microfarads (μF) or picofarads (pf). As long as your reading is within the tolerance of the marked value of the capacitor, you know the part is good.

If your multimeter is not equipped with an internal capacitor checker, you could use the resistance ranges to approximate the quality of a capacitor. This type of check provides a "quick and dirty" judgment of whether the capacitor is good or bad. The principle behind this is simple—all ohmmeter ranges use an internal battery to supply current for the component under test. When that current is supplied to a working capacitor as shown in Fig. 3-15, it will charge the capacitor. Charge accumulates as the ohmmeter is left connected, and can be seen as changing resistance on the ohmmeter display.

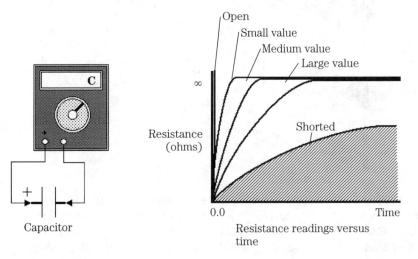

3-15 Measuring capacitance with an ohmmeter.

When first connected across an ohmmeter, the capacitor will draw a relatively large amount of current—this indicates low resistance. As the capacitor charges, it draws less and less current, so resistance appears to increase. Ideally, a fully charged capacitor draws no current, so your resistance reading should climb to infinity. When a capacitor behaves this way, it is probably good.

Understand that you are not actually measuring resistance or capacitance here, but only the profile of the charging characteristic of a capacitor. If the capacitor is extremely small, or is open circuited, it will not charge substantially, so it will instantly read infinity. If a capacitor is partially (or totally) short-circuited, it will not hold a charge, so you might read zero ohms (or resistance can climb to some level below infinity and remain there). In either case, the capacitor is probably defective. If you doubt your readings, check several other capacitors of the same value and compare readings. Be sure to make this test on a moderate to high resistance scale. A low resistance scale can charge to infinity too quickly for a clear reading.

Semiconductor checks

Many multimeters offer a semiconductor junction checker for diodes and transistors. Meters equipped with a diode range in their resistance function can be used to measure the static resistance of most common diodes in their forward or reverse-biased conditions as shown in Fig. 3-16.

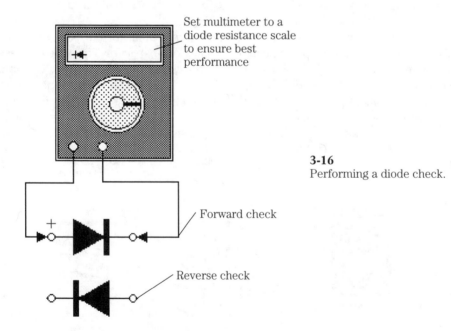

Set multimeter to a diode resistance scale to ensure best performance

Forward check

Reverse check

3-16
Performing a diode check.

Select the diode range from your meter resistance function and place test leads across the diode in the forward direction. A working silicon diode should exhibit a resistance between about 450 and 700 Ω that will read directly on your meter. Reverse your test leads to reverse-bias the diode. Because a working diode should not conduct at all in the reverse direction, you should read infinite resistance.

A shorted diode will exhibit a very low resistance in the forward and reverse-bias directions. This symptom indicates a shorted semiconductor junction. Be certain that at least one of the two diode leads is removed from the circuit before testing. This will prevent its interconnections with other components from causing a faulty reading. An opened diode will exhibit very high resistance (usually infinity) in both its forward and reverse directions. In this case, the semiconductor junction is open-circuited. If you feel unsure how to interpret your measurements, test several other comparable diodes and compare readings.

Transistors can be checked in several ways. Some multimeters feature a built-in transistor checker that measures transistor gain (or h_{fe}) directly. If your meter offers a transistor checker, insert your transistor into the test fixture on the meter face in its correct lead orientation (emitter, base, and collector). Manufacturer's specifications can tell you whether a gain reading is correct for a particular part. A low (or zero) reading indicates a shorted transistor, and a high (or infinite) reading suggests an open-circuited transistor.

Your meter's diode checking feature also can be used to check a bipolar transistor base-emitter and base-collector junctions as shown in Fig. 3-17. Each junction acts just like a diode junction. Test one junction at a time. Set your multimeter to its diode range, then place its test leads across the base-collector junction. If your transistor is NPN, place the positive test lead at the base. This arrangement should forward bias the base-collector junction and cause a normal amount of diode resistance. Reverse your test leads across the base-collector junction. The transistor should now be reverse-biased and show infinite resistance. Repeat this procedure for the base-emitter junction.

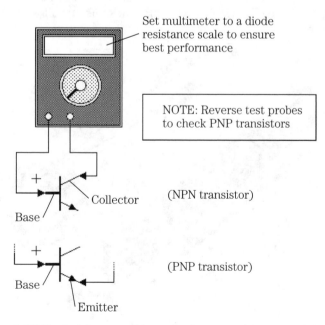

Set multimeter to a diode resistance scale to ensure best performance

NOTE: Reverse test probes to check PNP transistors

(NPN transistor)

(PNP transistor)

3-17 Performing a transistor check.

If your transistor is PNP, your test lead placement must be reversed. For example, a forward-biased junction in an NPN transistor is reverse-biased in a PNP transistor. You can refer to manufacturer's specification sheets to determine which leads in the transistor are the base, emitter, and collector.

As a final check, measure the resistance from emitter to collector. Note that you should read infinite resistance in both directions. Although this is not a diode junction, short circuits can develop during a transistor failure that might not appear across normal junctions. Replace any diode with an open or shorted junction, or a short from emitter to collector.

Logic probes

The problem with most multimeters is that they do not work well with digital logic circuits. A multimeter can certainly measure whether a logic voltage is on or off, but if that logic level changes quickly, a dc voltmeter function cannot track it properly. Logic probes provide a fast and easy means of detecting steady-state or alternating logic levels. Some logic probes can detect logic pulses faster than 50 MHz.

Logic probes are rather simple-looking devices as shown in Fig. 3-18. A probe can be powered from its own internal battery, or from the circuit under test. Connect the probe ground lead to a convenient circuit ground. If a probe is powered from the circuit under test, attach its power lead to a logic supply voltage in the circuit. A small panel on the probe body holds several LED indicators and a switch that allows the probe to work with two common logic families: TTL (transistor-transistor logic) and CMOS (complementary metal-oxide semiconductor). You might find TTL and CMOS devices mixed into the same circuit, but one family will usually dominate.

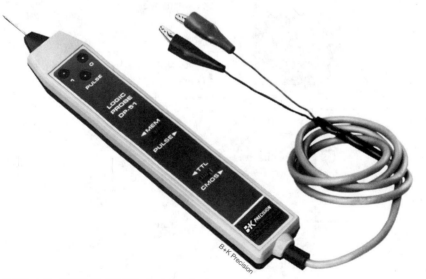

3-18 A B+K DP-51 50 MHz logic probe.

When the metal probe tip is touched to an IC lead, its logic state is displayed on one of the three LED indicators as shown in Table 3-1. Typical choices are LOW, HIGH, and PULSE (or CLOCK), but other indications can be presented as well depending on the sophistication of the probe. Logic probes are most useful for troubleshooting working logic circuits where logic levels and clock signals must be determined quickly and accurately.

Table 3-1. Typical logic probe display patterns

Input signal	HIGH LED	LOW LED	PULSE LED
Logic 1 (TTL or CMOS)	On	Off	Off
Logic 0 (TTL or CMOS)	Off	On	Off
Bad logic level or open circuit	Off	Off	Off
Square wave (<200 kHz)	On	On	Blink
Square wave (>200 kHz)	On/Off	On/Off	Blink
Narrow high pulse	Off	On/Off	Blink
Narrow low pulse	On/Off	Off	Blink

Oscilloscopes

Oscilloscopes offer a great advantage over multimeters and logic probes. Instead of reading signals in numbers or with lighted indicators, an oscilloscope will show voltage versus time on a visual display. Not only can you observe ac and dc voltages, but it enables you to watch digital voltage levels or other unusual signals occur in real time. If you have used an oscilloscope in the past, you know just how useful it can be. Oscilloscopes such as the one shown in Fig. 3-19 might appear somewhat overwhelming at first, but many of their operations work the same way regardless of what model you are working with.

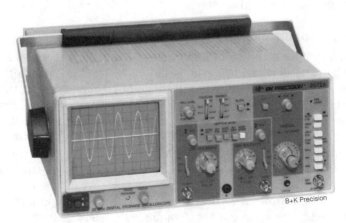

3-19 A B+K 2522A 20 MHz digital oscilloscope.

Controls

In spite of their wide variations of features and complexity, most controls are common to every oscilloscope. Controls fall into four categories: horizontal (time base) control, vertical signal control, housekeeping controls, and optional (enhanced) controls.

Housekeeping controls handle such things as oscilloscope power, trace intensity, graticule intensity, trace magnification, horizontal trace offset, vertical trace offset, and trace finder. Any control that effects the quality and visibility of a display.

Because an oscilloscope displays voltage versus time, adjusting either voltage or time settings will alter the display. Horizontal controls manipulate the left-to-right time appearance (sweep) of the voltage signal. Your oscilloscope master time base is adjusted using a TIME/DIV knob or button. This setting establishes the rate at which voltage signals are swept onto the screen. Smaller settings allow shorter events to be displayed more clearly, and vice versa. Remaining horizontal controls include a horizontal display-mode selector, sweep trigger selection and sensitivity, trigger coupling selection, and trigger source selection. Your particular oscilloscope might offer additional controls.

An adjustment to an oscilloscope voltage sensitivity also will alter your display. Vertical controls effect the deflection (up-to-down) appearance of your signal. An oscilloscope vertical sensitivity is controlled with the VOLTS/DIV knob. When sensitivity is increased (VOLTS/DIV becomes smaller), signals will appear larger vertically. Reducing sensitivity will make signals appear smaller vertically. Other vertical controls include coupling selection, vertical mode selection, and a display inverter switch.

Your oscilloscope might have any number of optional controls depending on its cost and complexity, but cursor and storage controls are some of the most common. Many scopes offer horizontal and vertical on-screen cursors to aid in the evaluation of waveforms. Panel controls allow each cursor to be moved around the screen. The distance between cursors is then converted to a corresponding voltage, time, or frequency value, and that number is displayed on the screen in appropriate units. Storage oscilloscopes allow a screen display to be held right on-screen, or in memory within the scope to be recalled on demand.

Oscilloscope specifications

Oscilloscopes have a variety of important specifications that you should be familiar with when choosing and using an oscilloscope. The first specification to know is *bandwidth*. Bandwidth represents the range of frequencies that the scope can work with. The bandwidth specification does not necessarily mean that all signals within that bandwidth can be displayed accurately. Bandwidth is usually rated from dc to some maximum frequency (often in megahertz—MHz). For example, an inexpensive oscilloscope might cover dc to 20 MHz, and a more expensive model might work up to 150 MHz or more. Good bandwidth is very expensive—more so than any other feature.

The *vertical deflection* (or vertical sensitivity) is another important specification. Deflection is listed as the minimum to maximum VOLTS/DIV settings that are offered, and the number of steps that are available within that range. A typical model might provide vertical sensitivity from 5 mV/DIV to 5 V/DIV broken down into 10 steps.

A *time base* (or sweep range) specification represents the minimum to maximum time base rates that an oscilloscope can produce, and the number of increments that are available. A range of 0.1 µs/DIV (microseconds per division) to 0.2 s/DIV in 20 steps is not unusual. You will typically find more time-base increments than sensitivity increments.

You must observe a maximum voltage input that can be applied to an oscilloscope input. A maximum voltage input of 400 V (dc or peak ac) is common for most basic models, but more sophisticated models can accept inputs better than 1,000 V. An oscilloscope input will present a load to whatever circuit or component it is placed across. This characteristic is called *input impedance*, and is usually expressed as a value of resistance and capacitance. To guarantee proper operation over the entire bandwidth of a model, select a probe with load characteristics to those of the oscilloscope. Most oscilloscopes have an input impedance of 1 MΩ with 10 to 50 pF of capacitance.

The accuracy of an oscilloscope represents the vertical and horizontal accuracy of the final CRT (cathode ray tube) display. In general, oscilloscopes are not as accurate as dedicated voltage or frequency meters. A typical model can provide ±3% accuracy, so a 1 V measurement can be displayed between 0.97 V to 1.03 V. Keep in mind that this does not consider human errors in reading the CRT marks (or *graticules*). However, because the strength of an oscilloscope is its ability to display complex and fast signals graphically, 3% accuracy is usually adequate.

Oscilloscope startup procedures

Before you begin taking measurements, get a clear stable trace (if not already visible). If a trace is not visible, make sure that any CRT screen storage modes are off, and that intensity is turned up at least 50%. Set triggering to its automatic mode and adjust the horizontal and vertical offset controls to the center of their ranges. Be sure to select an internal trigger source, then adjust the trigger level until a trace is visible. Vary your vertical offset if necessary to center the trace across the CRT.

If a trace is not yet visible, use the beam finder to reveal its location. A beam finder simply compresses the vertical and horizontal ranges. The compression forces a trace onto the display and gives you a rough idea of its relative position. After your trace is moved into position, adjust your focus and intensity controls to obtain a crisp, sharp trace. Keep intensity as low as possible to improve display accuracy, and preserve phosphors in the CRT.

Your oscilloscope probe must be calibrated before use. Calibration is a quick and straightforward operation that requires only a low-amplitude, low-frequency square wave. Many models have a built-in calibration signal generator (a 1 kHz, 300 mV square wave with a 50% duty cycle). Attach your probe to the desired input jack, then place it across the calibration signal. Adjust your horizontal (TIME/DIV) and vertical (VOLTS/DIV) controls so that one or two complete cycles are clearly shown on the CRT.

Observe the characteristics of your test signal as shown in Fig. 3-20. If the square wave corners appear rounded, there might not be enough probe capacitance (Cprobe). Spiked square wave corners suggest too much capacitance in the probe. Either way, the scope and probe are not matched properly. You must adjust the

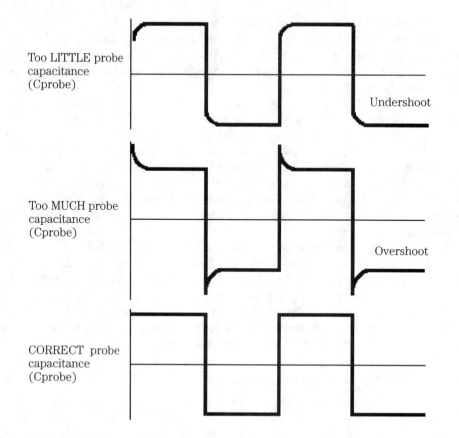

Too LITTLE probe capacitance (Cprobe)

Undershoot

Too MUCH probe capacitance (Cprobe)

Overshoot

CORRECT probe capacitance (Cprobe)

3-20 Oscilloscope calibration waveforms.

probe capacitance to establish a good electrical match—otherwise, signal distortion might result. Slowly adjust the variable capacitance on your probe until the corners of your calibration signal are as square as possible. If you cannot achieve a clean square wave, try a different probe.

Voltage measurements

The first step in any voltage measurement is to set your normal trace (or *baseline*) where you want it. Normally, a baseline is placed along the center of the graticule during start-up, but it could be placed anywhere so long as it is visible. To establish a baseline, switch your input coupling control to its ground position. This action disconnects the input signal and grounds the channel to ensure a zero reading. Adjust the vertical offset control to shift the baseline wherever the zero reading is to be. If you have no particular preference, simply center it in the CRT.

To measure dc, set your input coupling switch to its dc position, then adjust the VOLTS/DIV control to provide the desired amount of sensitivity. If you are unsure just which sensitivity is appropriate, start with a very low sensitivity (a large VOLTS/ DIV setting), then carefully increase the sensitivity (reduce the VOLTS/DIV setting) after your signal is connected. This action prevents a trace from simply jumping off the display when an unknown signal is first applied. If your signal does happen to leave the visible display, you could reduce sensitivity (increase the VOLTS/DIV setting) to make the trace visible again.

For example, suppose you were measuring a +5 Vdc power supply output. If VOLTS/DIV is set to 5 V/DIV, each major vertical division represents 5 V, so your +5 Vdc signal should appear 1 full division above your baseline (5 V/DIV × 1 division = 5 V) as shown in Fig. 3-21. At a VOLTS/DIV setting of 2 V/DIV, the same +5 V signal would now appear 2.5 divisions above your baseline (2 V/DIV × 2.5 divisions = 5 V). If your input signal were a negative voltage, the trace would appear below the baseline, but it would read the same way.

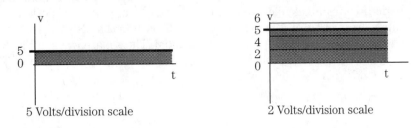

3-21 Measuring dc voltages with an oscilloscope.

You also can read ac signals directly from the oscilloscope. Switch your input coupling control to the ac position, then set a baseline just as you would for dc measurements. If you are unsure how to set the vertical sensitivity, start with a low sensitivity (a large VOLTS/DIV setting), then slowly increase the sensitivity (reduce the VOLTS/DIV scale) after a signal is connected. Keep in mind that ac voltage measurements on an oscilloscope will not match ac voltage readings on a multimeter. An oscilloscope displays instantaneous peak values for a waveform, and ac voltmeters measure in *rms* (root mean square) values. To convert an rms value to peak, multiply rms by 1.414. To convert a peak voltage reading to rms, divide peak by 1.414.

When actually measuring an ac signal, it might be necessary to adjust the oscilloscope trigger level control to obtain a stable (still) trace. As Fig. 3-22 shows, signal voltages can be measured directly from the display. For example, the sinusoidal waveform of Fig. 3-22 varies from −10 to +10 V. If scope sensitivity were set to 5 V/DIV, its peaks would be two divisions above and below the baseline. Because this is a peak measurement, an ac voltmeter would show the signal as peak/1.414 (10/1.414) or 7.07 V rms.

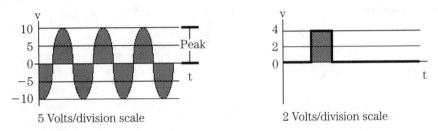

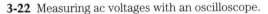

3-22 Measuring ac voltages with an oscilloscope.

Time and frequency measurements

An oscilloscope is perfect for measuring critical signal parameters such as pulse width, duty cycle, and frequency. The horizontal sensitivity (TIME/DIV) control comes into play with time and frequency measurements. Before making any measurements, you must first obtain a clear baseline as you would for voltage measurements. When a baseline is established and a signal is connected, adjust the TIME/DIV control to display one or two complete cycles of the signal.

Figure 3-23 shows two typical period measurements. With VOLTS/DIV set to 5 ms/DIV, the sinusoidal waveform repeats every 2 divisions. This wave represents a period of 10 ms (5 ms/DIV × 2 divisions). Because frequency is the simple reciprocal of the period, you can calculate frequency directly from period. A period of 10 ms represents a frequency of 100 Hz (1/10 ms). This process also works for square waves and other waveforms that are not sinusoidal. The square wave in Fig. 3-23 repeats every 4 divisions. At a TIME/DIV setting of 1 ms/DIV, its period is 4 ms. This period corresponds to a frequency of 250 Hz.

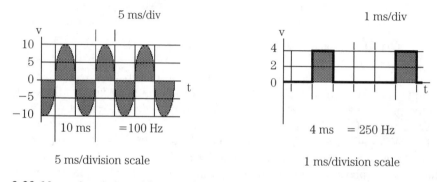

3-23 Measuring timing with an oscilloscope.

Instead of measuring the entire period of a pulse cycle, you also can read the time between any two points of interest. For the square wave in Fig. 3-23, you could read its pulse width to be 1 ms. You also could read the low portion of the cycle as a pulse width of 3 ms (added together for its total period of 4 ms). A signal *duty cycle* is simply the ratio of a signal ON time to its total period expressed as a percentage. For example, a square wave on for 2 ms and off for 2 ms would have a duty cycle of 50% [2 ms/(2 ms + 2 ms) × 100%]. For an on time of 1 ms and an off time of 3 ms, its duty cycle would be 25% [1 ms/(1 ms + 3 ms) × 100%]. **Use caution in duty-cycle measurements.**

4
CHAPTER

Service guidelines

Electronic troubleshooting is a strange pursuit; it is an activity that falls somewhere between art and science. Success in troubleshooting depends largely on a thorough, logical troubleshooting approach and the right type of test equipment, as well as an element of intuition and luck. This chapter shows you how to evaluate and determine printer problems, locate technical data, and present a series of service guidelines that can ease your work.

The troubleshooting cycle

Regardless of how complex your particular circuit or system might be, a reliable troubleshooting procedure can be broken down into four basic steps as shown in Fig. 4-1: (1) define your symptoms, (2) identify and isolate the potential source (or loca-

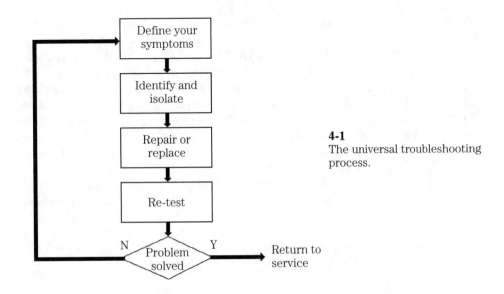

4-1
The universal troubleshooting process.

tion) of the problem, (3) replace or repair the suspected component or subassembly, and (4) re-test the system thoroughly to be sure that you have solved the problem. If you have not solved the problem, begin again from step 1. The procedure is a universal procedure you can use for any troubleshooting—not just for laser printers.

Define your symptoms

Sooner or later, a laser printer is going to break down. The problem might be as simple as a sticky gear or as complicated as an extensive electronic failure. However, before you open the toolbox, you must have a firm understanding of the symptoms. You must do more than to simply say, "It's busted." Think about its symptoms carefully. Ask yourself what is (or is not) happening. Consider when it is happening. If this installation is new, ask yourself if the computer is set up properly, or if the right cables are being used, or if DIP switches are set up correctly. If you have used your printer for a while, do you remember the last time you cleaned and lubricated it? Is the print light, dark, or completely missing? Is the paper advancing freely? By recognizing and understanding your symptoms, you will find it easier to trace a problem to the appropriate subsection or components.

Use your senses and write down as many symptoms as you can—whatever you smell, see, or hear. Writing symptoms might sound tedious now, but when you are up to your elbows in repair work, a written record of symptoms and circumstances will keep you focused on the task at hand. Writing symptoms is even more important if you are a novice troubleshooter.

Identify and isolate

Before you try to isolate a problem in the laser printer, first be sure that the printer is, in fact, causing the problem. In many circumstances, printer problems will be obvious, but there are some situations that might appear ambiguous (no print with power on, erratic printing, not enough contrast, etc.). Always remember that a printer is just a subsection of a larger system including your computer, laser printer, and interconnecting cable. Especially in new installations, a computer failure, software incompatibility, or cable problem might be causing your symptoms.

An easy application of the universal troubleshooting procedure follows. Once you have carefully identified your symptoms, isolate the printer. You can isolate a printer by removing it from its communication cable. You can replace it by testing it on another computer system with a working printer (one that you know is working well). A friend or colleague might let you test your printer on their computer system. Because various computers can be set up to communicate in different fashions, you might have to alter the internal settings of your printer to match those of the working printer. When printing from an operating system such as Microsoft Windows, you might have to select and configure a new printer driver to support your printer. If your printer exhibits the same symptoms on another computer, there is an excellent chance that the problem is within the printer. You can then proceed with specific troubleshooting procedures. If, however, those symptoms disappear and your printer works properly, you should suspect a problem in your computer, software configuration, DIP switch settings, printer driver, or interconnecting cable.

Another test is to try a working printer on your computer system. As before, you might need to select another printer driver to operate the working printer properly on your existing system. If another printer works properly, it verifies that the computer, software configuration, and cable are intact. If a working printer fails to work on your system, check the computer communication interface, software settings, and interconnecting cable. Complete this check in addition to testing your questionable printer on another system.

When you are confident that the printer is at fault, you can begin to identify any possible problem areas. Start at the subsection level. You might recall from chapter 1 that a laser printer consists of several major subsections. Your printer fault will be located in at least one of these subsections or subassemblies. The troubleshooting procedures of chapters 6 through 9 will aid you in deciding which subsections are at fault. Once you have identified a potential problem, you can begin the actual repair process. In many cases, your repair will involve replacing a defective subassembly. For skilled technicians, the repair might include tracking a defect to the component level.

Repair or replace

Once you have an understanding of what is wrong and where to look, you might begin the actual repair procedures that you think will correct the symptoms. Some procedures require only simple adjustments or cleaning, and others might require the exchange of electrical or mechanical parts. **All procedures are important and should be followed very carefully.**

Parts are usually classified as *components* or *subassemblies*. A component part is the smallest possible individual part that you can work with. Components can serve many different purposes in a printer. Resistors, capacitors, gears, belts, motors, and integrated circuits are just a few types of component parts. Usually, components contain no serviceable parts—the components themselves must be replaced. A subassembly is composed of a variety of individual components. Unlike components, a complete subassembly serves a single, specific purpose in a printer, but it too can be repaired by locating and replacing any faulty components. Repairing a defective subassembly simply by installing a new one in the printer is certainly an acceptable solution.

All technicians must make the cost/performance tradeoff when performing a repair. Component parts are much less expensive than subassemblies, but components are often specialized and can be difficult to get. You also might need test equipment and time to troubleshoot to the component level. Replacing subassemblies is faster and easier than tracing component faults, even though assemblies are more expensive. Subassembly service makes good sense for individuals who lack the time, experience, or test equipment to worry about component-level faults.

Replacement electronic components might often be purchased from several different sources, but keep in mind that many mechanical parts and fittings might only be available through the manufacturer or distributor. Many of the mail-order companies listed at the end of this book will send you their complete catalogs or product listings at your request. Going to the manufacturer for subassemblies or components is often somewhat of a calculated risk—they might do business only with their affiliated service centers, or refuse to sell parts directly to consumers. If you find a manufacturer willing to sell you parts, you must often know the manufacturer's exact part number or code. Remember that many manufacturers are ill equipped to deal with consumers directly, so be patient and be prepared to make several different calls.

During a repair, you might reach a roadblock that requires you to leave the printer for a day or two (or longer). The delay is typical when you have diagnosed a failure and are waiting for parts. Make it a point to reassemble the printer as much as possible before leaving it. Place any loose parts into plastic bags and seal them shut. Reassembly will prevent a playful pet, curious child, or well-meaning spouse from accidentally misplacing or discarding parts while the printer sits on your workbench. Making loose parts secure is twice as important if your workspace is in a well-traveled or family area. You also will remember how to put it back together later on. Make notes to remind yourself what parts go where.

Re-test

When repair is complete, carefully reassemble the laser printer and test it before connecting it to a computer. Run a thorough self-test to check printer operation. The self-test checks the image-formation system, paper pickup and registration, fusing assembly, power supply, and much of the printer electronics. If symptoms persist, you will have to re-evaluate them and narrow the problem to another part. If normal operation is restored (or significantly improved), test the printer with a computer and interconnecting cable. When you can verify that your symptoms have stopped during actual operation, the printer can be returned to service.

Do not be discouraged if the printer still malfunctions. Simply walk away, clear your head, and start again by defining your symptoms. Never continue with a repair if you are tired or frustrated—tomorrow is another day. For technicians troubleshooting to the component level, also realize that there might be more than one bad component to deal with. Remember that a laser printer is just a collection of assemblies, and each assembly is a collection of components. Normally, everything works together, but when one part fails, it might cause one or more interconnected parts to fail as well. Be prepared to make several repair attempts before the printer is repaired completely.

Gathering technical data

Technical information is perhaps your most valuable tool in tackling a printer repair. Just how much information you actually need will depend on the particular problems you are facing. Simple adjustments and cleaning might be accomplished with little or no specific technical information (except your own observations and common sense judgment), but complex electronic troubleshooting might require a complete set of schematics. Parts lists will be needed to order new mechanical components and all types of subassemblies. More intricate repair procedures generally need more comprehensive technical literature. Luckily, there are some avenues of information.

Your user's manual is always a good place for basic printer information. A user's manual describes how to set up and operate the printer, outlines its important specifications and communication interface, and points out its major assemblies and controls. If you are unfamiliar with the printer or unaccustomed to changing its configurations, a user's manual can keep you out of trouble. Some user's manuals also present a short selection of very basic troubleshooting and maintenance procedures, but these are almost always related to the printer setup and operation—not to its internal circuitry or mechanics.

You can find technical information on many individual components on data sheets published by the component manufacturer. For example, if you want a pin diagram of an IC manufactured by Motorola, you could refer to a Motorola data book containing information on that particular component. The data book tells you what the part is, what it does, what purpose each pin performs, and what its electrical specifications are. Although data books bear no direct relationship to your particular printer, they can give you much insight on the purpose and functions of individual components.

However, if you intend to pursue detailed electronic repairs, you will need a set of schematics. A complete set of schematics can quickly and efficiently guide you through even the most complicated printer. If you are working on an older printer, there might be a complete documentation package published by Howard W. Sams & Co. Their comprehensive *Sams Photofact* series has long been an indispensable part of the electronic service industry. The address and phone number for Howard W. Sams & Co. are listed in appendix C. A manufacturer's maintenance manual also offers parts lists and mechanical diagrams that clarify how the printer is assembled.

Your printer manufacturer can be a key source of information, but not all manufacturers are willing to sell technical information to individuals or private organizations. Start by checking directly with the manufacturer. Their phone number is usually listed somewhere in the user's manual. If no user manual is available, you can probably find the manufacturer in appendix C. You can try to contact their technical literature, parts order, or service departments to order a service or repair manual. Service information can be expensive (as much as $50 or more) so be prepared.

If you cannot get satisfaction from the manufacturer, check with a local dealer (not a retail store) that sells for that manufacturer. The *Yellow Pages* of your local telephone book can give you good leads. A reputable dealer can get parts and technical information that you cannot. Finally, try contacting a service organization that repairs your type of printer. They might be willing to order a copy for you, but some organizations prefer that you bring the printer in for their repair services.

Electricity hazards

No matter how harmless your printer might appear, always remember that potential shock hazards exist. Once the printer is disassembled, there can be several locations where live ac voltage is exposed and easily accessible. Domestic electronic equipment operates on 120 Vac at 60 Hz. Some European countries use 240 Vac at 50 Hz. When voltage of this level establishes a path through your body, it causes a flow of current that might be large enough to stop your heart. Because it only takes about 100 mA (milliamperes) to trigger cardiac arrest, and a typical printer fuse is rated for 1 or 2 A, fuses and circuit breakers will NOT protect you. The high voltage available in laser printers also present serious shock hazards.

The resistance of your skin limits the flow of current through the body. According to Ohm's law, any voltage, current flow increases as resistance drops (and vice versa). Dry skin exhibits a high resistance of several hundred thousand ohms; and moist, cut, or wet skin can drop to only several hundred ohms. This means that even comparatively low voltages can produce a shock if your skin resistance is low enough. Some examples help to demonstrate this action.

Suppose your hands contact a live 120 Vac circuit. If your skin is dry (say 120 kΩ), you would experience an electrical shock of 1 mA (120 Vac/120,000 Ω). The result would be harmless—probably a brief, tingling sensation. After a hard day's work, perspiration could decrease skin resistance (perhaps to 12 kΩ). This would allow a far more substantial shock of 10 mA (120 Vac/12,000 Ω). At that level, the shock can paralyze the victim and make it difficult or impossible to let go of the "live" conductors. A burn (perhaps serious) could result at the points of contact, but it probably would not be fatal. Consider a worker whose hands or clothing are wet. The effective skin resistance can drop very low (1.2 kΩ for example). At 120 V, the resulting shock of 100 mA (120 Vac/1,200 Ω) would often be instantly fatal unless immediate CPR is administered. **Use EXTREME caution whenever working around circuitry with live power exposed.**

Electrostatic printers use high-voltage power supplies that are even more dangerous. Most can produce voltage easily exceeding –2,000 Vdc. Based on the examples you just read, even dry skin at 200,000 Ω could receive a paralyzing shock of 10 mA (2,000 Vdc/200 kΩ). Fortunately, high-voltage power supplies are not designed to allow significant current to flow, but serious burns can be delivered with ease. Not only is there a great risk of injury, but normal test probes (such as multimeter test leads) only provide insulation to about 600 V. Testing high voltages with standard test leads could electrocute you right through the lead insulation! Be sure to use specially designed high-voltage probes when measuring high-voltage points. Take the following steps to protect yourself from injury:

1. **Keep the printer unplugged (not just turned off) as much as possible during disassembly and repair.** When you must perform a service procedure that requires power to be applied, plug in the printer just long enough to perform your procedure, then unplug it again. This makes the printer safer for you, as well as your spouse, child, and pets that might happen along. For added safety, plug in your printer through an *isolation transformer* (Fig. 4-2).

2. **Whenever you must work on a power supply, wear rubber gloves.** The gloves will insulate your hands just like insulation on a wire. You might think that rubber gloves are inconvenient and uncomfortable, but they are far better than the inconvenience and discomfort of an electric shock. Make it a point to wear a long-sleeved shirt with sleeves rolled down, which will insulate your forearms.

3. If you absolutely cannot wear rubber gloves for one reason or another, remove all metal jewelry and work with one hand behind your back. **The metals in your jewelry are excellent conductors.** Should your ring or watchband hook onto a live ac line, jewelry can conduct current directly to your skin. By keeping one hand behind your back, you cannot

4-2
A B+K 1604 isolation transformer.

grasp both ends of a live ac line to complete a strong current path through your heart.

4. Inspect your test probes carefully before testing high-voltage circuitry. **Standard "off-the-shelf" probes do not necessarily have the insulating properties (called *dielectric strength*) to protect you.** If you must make powered tests on a high-voltage circuit, be sure to use test leads that offer sufficient protection.

5. **Work dry!** Do not work with wet hands or clothing. Do not work in wet or damp environments. Make sure that any available fire extinguishing equipment is suitable for electrical fires.

6. Treat electricity with the proper respect. Whenever electronic circuitry is exposed (especially power supply circuitry), a shock hazard exists. Remember that it is the flow of current through your body, not the voltage potential, that can injure you. **Insulate yourself as much as possible from any exposed wiring.**

Static electricity

Another troubleshooting hazard can come from static voltages accumulated on your body or tools. If you have ever walked across a carpeted floor on a dry winter day, you have probably experienced the effects of *ESD* (electrostatic discharge) first hand while reaching for a metal object. Under the right conditions, your body can accumulate static charge potentials greater than 20,000 V. When you provide a conductive path for electrons to flow, that built-up charge rushes away from your body at the point closest to the object. The result is often a brief, stinging shock. Such a jolt can be startling and annoying, but is generally harmless to people. Semiconductor devices, however, are highly susceptible to damage from ESD while you handle or replace circuit boards and components. This section introduces you to static electricity, and shows you how to prevent ESD damage during your repairs.

Static formation

When two dissimilar materials are rubbed together, the force of friction causes electrons to move from one material to another. The excess (or lack) of electrons cause a charge to develop on each material. Because electrons are not flowing, there is no current, so the charge is said to be *static*. However, the charge does exhibit a voltage potential. As materials continue to rub together, their charges increase—sometimes to potentials of thousands of volts.

In a human, static charges can be developed by normal everyday activities such as walking on a carpet. Friction between the carpet and shoe soles cause opposing charges to be developed. The charge on the shoe induces an equal (but opposite) charge in your body, which acts as a capacitor. Sliding across a vinyl car seat, pulling a sweater on or off, or taking clothes out of a dryer are just some of the ways that a static charge can appear in the body.

Device damage

ESD poses a serious threat to virtually all modern semiconductor devices. Huge static voltages that build up in the environment (or in your body) can find their way into all types of advanced ICs. If that happens, the result for the component can be catastrophic. Static discharge can damage bipolar transistors, TTL devices, ECL (emitter-coupled logic)devices, operational amplifiers, SCRs (silicon-controlled rectifiers), and JFETs (junction field-effect transistors). Certainly the most susceptible components are those fabricated using MOS (metal-oxide semiconductor) technology. A typical MOS transistor is shown in Fig. 4-3.

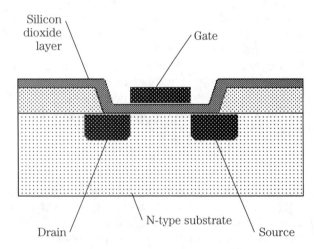

4-3
Diagram of a typical MOS transistor.

MOS devices (PMOS, NMOS, HMOS, CMOS, etc.) have become the cornerstone of high-performance ICs as memories, high-speed logic, microprocessors, and other advanced digital components. MOS technologies offer high-speed, high component density, and low power consumption. Typical MOS ICs can easily cram over 1 million transistors onto a single IC. Every part of the transistor must be made continually smaller to keep pace with the demands for higher levels of integration. As each part of the transistor shrinks, however, breakdown voltages drop, and ESD damage problems escalate.

A typical MOS transistor breakdown is shown in Fig. 4-4. Notice the areas of positive and negative semiconductor material that forms its three-terminals: source, gate, and drain. The *gate* is isolated from other parts of the transistor by a thin film of silicon dioxide (sometimes called the *oxide layer*). Unfortunately, this layer is extremely thin, and it can be easily overcome by high voltages like those from static discharges. Once this happens, the oxide layer is punctured. This renders the entire transistor (and the whole IC) permanently defective.

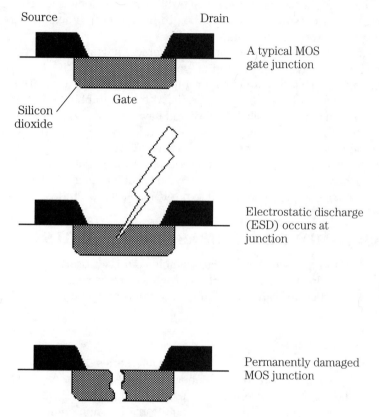

4-4 The effect of MOS breakdown.

Controlling static electricity

Do not underestimate the importance of static control during your printer repairs. Without realizing it, you could destroy a new IC or circuit assembly before you even have a chance to install it—and you would never even know that static damage has occurred. All it takes is the careless touch of a charged hand, tool, or piece of clothing. Take the necessary steps to ensure the safe handling and replacement of your sensitive (and expensive) electronics.

One way to control static is to keep charges away from boards and ICs. This static control is often a part of device packaging and the shipping container. ICs are usually packed in a specially-made conductive foam. Carbon granules are compounded right into the polyethylene foam to achieve conductivity of about 3,000 Ω/cm (ohms per centimeter). Foam prevents bending of IC leads, absorbs vibrations and shocks, and its conductivity helps to keep every IC lead at the same potential (also called *equipotential bonding*). Conductive foam is reusable, so you can insert ICs for safe keeping, then remove them as needed. You can purchase conductive foam from almost any electronics retail store.

Circuit boards are normally held in conductive plastic bags that dissipate static charges before damage can occur. Antistatic bags are made up of different layers— each with varying amounts of conductivity. The bag acts as a *Faraday cage* for the device it contains. Electrons from an ESD will dissipate along the surface layers of the bag instead of passing through the bag to its contents. Bags also are available through many electronics retail outlets.

Whenever you work with sensitive electronics, it is a good idea to dissipate charges that might have accumulated on your body. A conductive fabric wrist strap that is soundly connected to an earth ground will bleed away all charges from your skin. Avoid grabbing hold of a ground directly. Although grabbing the ground will discharge you, it can result in a sizable jolt if you have picked up a large charge.

Remember to make careful use of your static controls. Keep ICs and circuit boards in their antistatic containers at all times. *Never* place parts onto synthetic materials (such as plastic cabinets or fabric coverings) that could hold a charge. Handle static sensitive parts carefully and avoid touching their metal pins if possible. Be sure to use a wrist strap connected to a reliable earth ground.

Reassembly and disassembly hints

Sooner or later, you will have to disassemble your laser printer to some extent to perform your repair. Although the actual process of disassembly and reassembly is usually pretty straightforward, there are some important points for you to keep in mind during your procedures.

Housing disassembly

Most laser printer enclosures are designed as a series of covers that are latched together. By opening the top cover, some assemblies should be exposed. This gives you quick and easy access to replaceable parts (such as the EP cartridge). In many cases, however, you will have to remove additional housing components to access major assemblies (such as power supplies, the fuser, and main logic). Examine your enclosure very carefully before beginning the disassembly. Some enclosures are held together with simple screws in an obvious, easily accessible fashion. Other types of enclosures use unusual screw patterns such as Spline or Torx. They also can incorporate cleverly hidden internal clips that latch the enclosures together and provide a *seamless* appearance. Seamless housings might need special tools to disengage these internal latches before housings can be separated. You might like to get another person's assistance when disassembling this type of housing.

Electromechanical disassembly

Safety is critically important whenever you are working with electronic circuitry, **so be sure to UNPLUG THE PRINTER before starting any work.** Plug it in only long enough to follow your particular troubleshooting and testing procedures, then unplug it again.

Laser printers contain a wide array of electrical connectors handling everything from ac line voltage to laser control signals. During disassembly, you might have to remove one or more connectors to free a circuit board or other subassembly. *Never* remove a connector by yanking on its wires—many connector shells use keys or latches to hold them in place. Always remove a connector by holding its shell. Take careful note of each connector location and orientation. Some connectors are keyed so they can only be reinserted in their proper orientation, but other types of connectors might not be this foolproof. Make a sketch of where everything goes before taking things apart.

Take careful note of physical parts as well, especially when you must disassemble complex drive trains of gears or pulleys. It will help you tremendously when it comes time to reassemble the system. Mark your parts before disassembly with an indelible felt-tip marker. Feel free to use any kind of markings that are clear to you, but marks should show how each part is mounted in relation to its adjacent parts. Fig. 4-5 is just one simple example.

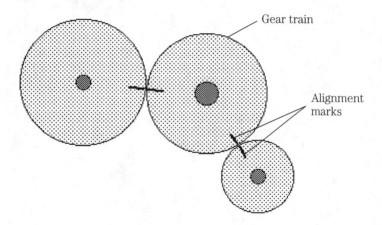

4-5 Making alignment marks before disassembly.

Reassembly

Whenever you must replace ICs on a through-hole printed circuit board, always solder an IC holder in its place, then plug the replacement IC into its holder. Printed circuit boards (especially complex boards) are very delicate, and printed traces can be damaged by excessive or repeated heating. If you install an IC holder in its place, you will never have to desolder those points again. To replace that IC in the future, just unplug it and install a new one. Keep in mind there might not be enough room to install an IC socket if the board is located close to an enclosure of another assembly. Use your best judgment to decide if there is enough room for a socket.

Always double check your connector locations and orientations before applying power. If a connector is engaged backward or is skipping pins, your circuits can be seriously damaged. If you have made orientation marks on the connectors before disassembly, they should be a snap to install properly.

Metal shields or shrouds are often added to limit *RF* (radio-frequency) interference between circuits. Switching-regulated power supplies and high-speed devices such as microprocessors are often shielded thoroughly. This prevents noise generated in one circuit from causing false signals in another circuit. If you have ever seen or heard radio or television reception near a computer, then you have probably witnessed the effects of RF noise. Because a printer uses many of the same electronic components that a computer does, it too can generate noise. Be certain that all RF noise shielding is installed and secured properly. You can add metal or plastic guards to protect physical parts such as drive trains. Be sure you replace all protective covers before re-testing your laser printer.

5
CHAPTER

Electrophotographic technology

Electrophotographic (EP) printers are different from other types of printers that you might be familiar with (such as dot matrix or ink jet printers). Those conventional printers develop dots as a one-step process using impact, heat, or ink. EP printers (Fig. 5-1) are not nearly as simple. EP images are formed by a complex and delicate interaction of light, static electricity, chemistry, pressure, and heat—all guided by a sophisticated electronic control package (ECP). This chapter details the background of EP technology, and explains the operation of the image-formation system of your laser printer.

Hewlett-Packard Co.

5-1
A Hewlett-Packard LaserJet IV printer.

The electrophotographic approach

As a result of the interaction of elements, EP printing is accomplished through a *process* rather than through a print head. The collection of components that performs the EP printing process is called an *IFS* (image-formation system). An IFS is made up of eight distinctive areas (see Fig. 5-2): a photosensitive drum (14), cleaning blade, erasure lamp (3), primary corona (4), writing mechanism (5 and 6), toner, transfer corona (13), and fusing rollers (18 and 19). Each of these parts play an important role in the proper operation of an IFS.

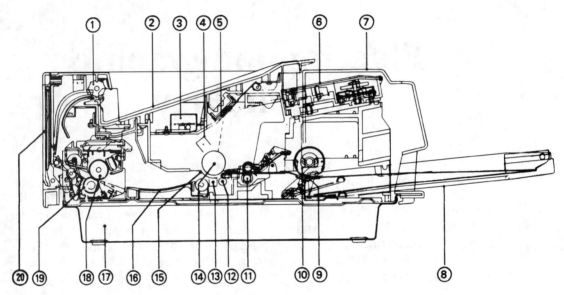

1. Delivery assembly
2. Face-down tray
3. Erase lamp assembly
4. Primary corona
5. Beam-to-drum mirror
6. Laser/scanning assembly
7. Main body covers
8. Paper tray
9. Separation pad
10. Feed roller assembly
11. Registration rollers
12. Transfer corona roller
13. Transfer corona assembly
14. Photosensitive EP drum
15. EP drum protective shield
16. Feed guide assembly
17. Lower main body
18. Upper fusing roller
19. Lower pressure roller
20. Face-up output tray (closed)

5-2 Cross-sectional diagram of a laser printer. Hewlett-Packard Co.

A photosensitive drum, such as the one shown in Fig. 5-3, is considered the heart of any IFS. An extruded aluminum cylinder is coated with a nontoxic organic compound that exhibits photoconductive properties. That is, the coating will conduct electricity when exposed to light. The aluminum base cylinder is connected to ground of the high-voltage power supply.

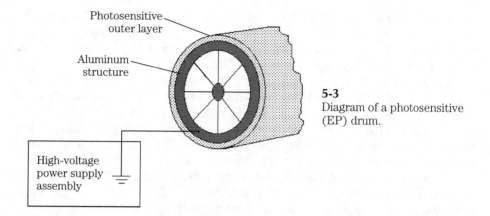

5-3
Diagram of a photosensitive (EP) drum.

It is the drum that actually receives an image from a *writing mechanism*, develops the image with toner, then transfers the developed image to paper. Although you might think that this constitutes a print head because it delivers an image to paper, the image is not yet permanent—other operations must be performed by the IFS. Complete image development is a six-step process that involves all eight IFS components: cleaning, charging, writing, developing, transfer, and fusing. To understand the IFS, you should know each of these steps in detail.

Cleaning

Before a new printing cycle can begin, the photosensitive drum must be physically cleaned and electrically erased. Cleaning might sound like an unimportant step, but not even the best drum will transfer every microscopic granule of toner to a page every time. A rubber cleaning blade is applied across the entire length of the drum to gently scrape away any residual toner that might remain from a previous image. If residual toner were not cleaned, it could adhere to subsequent pages and appear as random black *speckles*. Toner that is removed from the drum is deposited into a debris cavity as shown in Fig. 5-4. Keep in mind that cleaning must be accomplished without scratching or nicking the drum. Any damage to the photosensitive drum sur-

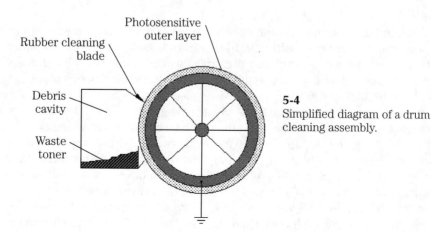

5-4
Simplified diagram of a drum cleaning assembly.

face would become a permanent mark that would appear on every subsequent page. Some EP printer designs actually return scrap toner back to the toner supply for reuse. This kind of recycling technique can extend the life of your electrophotographic (EP) cartridge and eliminate the need for a large debris cavity.

Images are written to a drum surface as horizontal rows of electrical charges that correspond to the image being printed. A dot of light causes a relatively positive charge at that point. The dot of light corresponds to a visual dot in the completed image. Absence of light allows a relatively negative charge to remain and no dots are generated. The charges caused by light must be removed before any new images can be written—otherwise images would overwrite and superimpose on one another.

A series of *erase lamps* are placed near the drum surface. Their light is filtered to allow only effective wavelengths to pass. Erase light bleeds away any charges along the drum. Charges are carried to ground through the aluminum cylinder as shown in Fig. 5-5. After erasure, the drum surface is completely neutral—it contains no charges at all.

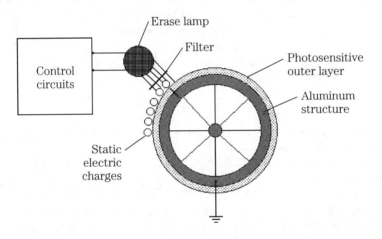

5-5 Discharging the EP drum with an erase lamp.

Charging

The neutral drum surface is no longer receptive to light from the writing mechanism. New images cannot be written until the drum is charged again. To charge (or *condition*) the drum, a uniform electrical charge must be applied evenly across its entire surface. Surface charging is accomplished by applying a tremendous negative voltage (often more than –6,000 V) to a solid wire called a *primary corona* located close to the drum. Because the drum and high-voltage power supply share the same ground, an electrical field is established between the corona wire and drum as in Fig. 5-6.

At low voltages, the air gap between a corona wire and drum would act as an insulator. With thousands of volts of potential, however, the insulating strength of air breaks down and an electric *corona* forms. A corona ionizes any air molecules surrounding the wire, so negative charges migrate to the drum surface.

The trouble with ionized gas is that it exhibits a very low resistance to current flow. Once a corona is established, there is essentially a short circuit between the

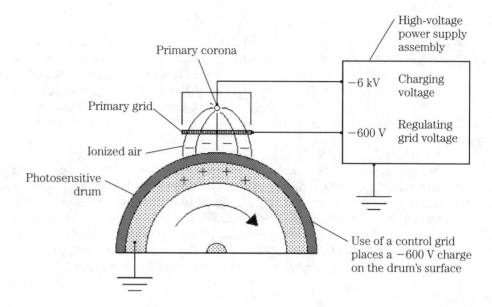

5-6 Conditioning the drum with the primary corona.

wire and drum. The short circuit is not good for a high-voltage power supply. A *primary grid* (part of the primary corona assembly) is added between the wire and drum. By applying a negative voltage to the grid, charging voltage and current to the drum can be carefully regulated. This *regulating grid voltage* (often –600 to –1,000 V) sets the charge level actually applied to the drum that is typically equal to the regulating voltage (–600 to –1,000 V). The drum is now ready to receive a new image.

Writing

To form a latent image on a drum surface, the uniform charge that has conditioned the drum must be discharged in the precise points where images are to be produced. Images are written using light. Any points on the drum exposed to light will discharge to a very low level (about –100 V), and any areas left unexposed retain their conditioning charge (–600 to –1,000 V). The device that produces and directs light to the drum surface is called a *writing mechanism.*

Because images are formed as a series of individual dots, a higher number of dots per area allows finer resolution (and higher quality) of the image. For example, suppose a writing mechanism can place 300 dots per inch on a single horizontal line on the drum, and the drum can rotate in increments of ⅟₃₀₀ of an inch. Using these specifications, your printer can develop images with a resolution of 300 × 300 dots per inch (dpi). Current EP printers are reaching 600 × 600 dpi.

Lasers have been traditionally used as writing mechanisms (thus the name "laser printer"), and are still used in many EP printer designs, but new printers are replacing lasers with bars of light-emitting diodes (LEDs) or arrays of liquid crystal shutters (LCSs) to direct light as needed. (Writing mechanisms are covered more extensively in this chapter.) Once an image has been written to a drum, that image must be developed.

Developing

Images written to the drum by laser or LED are initially invisible—merely an array of electrostatic charges on the drum surface. There are low charges where the light strikes, and high charges where the light skips. The latent image must be developed into a visible one before it can be transferred to paper. *Toner* is used for this purpose. Toner itself is an extremely fine powder of plastic resin and organic compounds bonded to iron particles. You can see the individual granules under extreme magnification of a microscope.

Toner is applied using a toner cylinder (or *developer roller*) as shown in Fig. 5-7. A toner cylinder is a long metal sleeve containing a permanent magnet. The cylinder is mounted inside the toner supply trough. When the cylinder turns, iron in the toner attracts it to the cylinder. Once attracted, toner acquires a negative static charge provided by the high-voltage power supply. This static charge level falls between the levels of the exposed and unexposed charge levels of the photosensitive drum. The drum levels can be from –200 to –500 V depending on the intensity control setting. A restricting blade limits toner on the cylinder to a single layer.

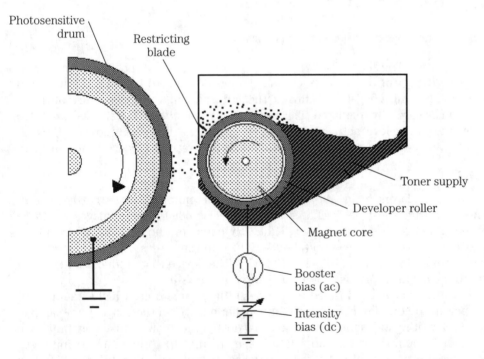

5-7 Applying toner to a charged EP drum.

Charged toner on the cylinder now rotates close to the exposed drum. Any points on the drum that are not exposed will have a strong negative charge. This charge repels toner that remains on the toner cylinder and is returned to the supply. Any points on the drum that are exposed now have a much lower charge than the toner particles. This charge attracts toner from the cylinder to corresponding points on the drum. Toner "fills-in" the latent image to form a visible (or *developed*) image.

Notice that an ac booster bias (more than 1,500 V_{pp}) is added in series to the dc intensity bias. The ac causes strong fluctuations in the toner charge level. As the ac signal goes positive, the intensity level increases to help toner particles overcome attraction of the permanent magnet of the cylinder. As the ac signal goes negative, intensity levels decrease to pull back any toner particles that might have falsely jumped to unexposed areas. This technique improves print density and image contrast. The developed image can now be applied to paper.

Transfer

At this point, the developed toner image on the drum must be transferred onto paper. Because toner is now attracted to the drum, it must be pried away by applying an even larger attractive charge to the page. A *transfer corona* wire charges the page as shown in Fig. 5-8. The theory behind the operation of a transfer corona is exactly the same as that for a primary corona—except that the potential is now positive. This places a powerful positive charge on paper, which attracts the negatively charged toner particles. Remember that this is not a perfect process—not all toner is transferred to paper. This is why a cleaning process is needed.

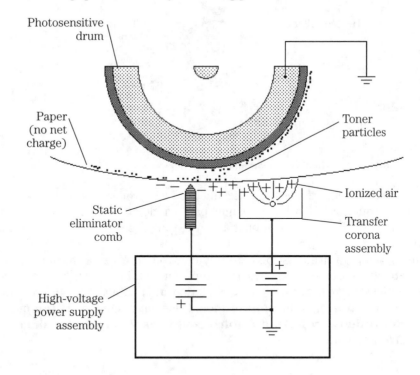

5-8 Transferring the developed image to paper.

Caution is needed here. Because the negatively charged drum and positively charged paper attract each other, it is possible that paper could wrap around the drum. Even though the small-diameter drum and natural stiffness of paper tend to prevent wrapping, a static charge eliminator (or *eliminator comb*) is included to counteract positive charges and remove the attractive force between paper and drum immediately after toner is transferred. The paper now has no net charge. The drum can be cleaned and prepared for a new image.

Fusing

Once the toner image has reached paper, it is only held to the page by gravity and weak electrostatic attraction—toner is still in its powder form. Toner must be fixed permanently (or *fused*) to the page before it can be handled. Fusing is accomplished with a heat and pressure assembly like the one shown in Fig. 5-9. A high-intensity quartz lamp heats a nonstick roller to about 180°C. Pressure is applied with a pliable rubber roller. When a developed page passes between these two rollers, heat from the top roller melts the toner, and pressure from the bottom roller squeezes molten toner into the paper fibers. In the fiber, the toner cools and adheres permanently. The finished page is then fed to an output tray. Note that both rollers are called *fusing rollers*, even though only the heated top roller actually fuses.

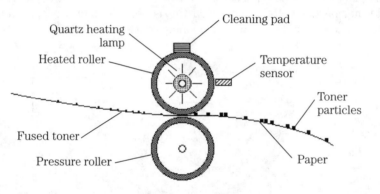

5-9 Fusing the transferred image to paper.

To prevent toner particles from sticking to a fusing roller, it is coated with a nonstick material such as Teflon™. A cleaning pad is added to wipe away any toner that might yet adhere. The pad also applies a thin coating of silicon oil to prevent further sticking.

Fusing temperature must be carefully controlled. Often a thermistor is used to regulate current through the quartz lamp to maintain a constant temperature. A snap-action thermal switch also is included as a safety interlock to prevent damage in case the lamp temperature should rise out of control. **If temperature is not controlled carefully, a failure could result in printer damage, or even a fire hazard.**

Writing mechanisms

After charging, the photosensitive drum contains a uniform electrostatic charge across its surface. To form a latent image, the drum must be discharged at any points that comprise the image. Light is used to discharge the drum as needed. Such a writing mechanism is shown in Fig. 5-10.

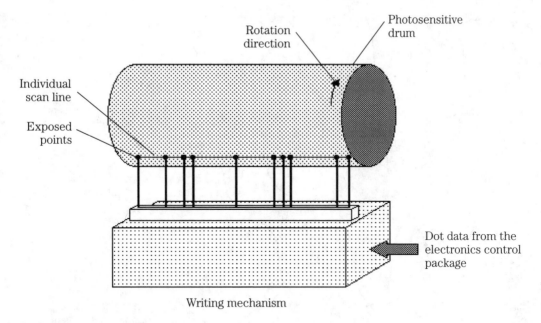

Rotation direction

Photosensitive drum

Individual scan line

Exposed points

Dot data from the electronics control package

Writing mechanism

5-10 Simplified diagram of a generic writing mechanism.

Images are scanned onto the drum one horizontal line at a time. A single pass across the drum is called a *trace* or *scan line*. Light is directed to any points along the scan line where dots are required. When a scan line is completed, the drum increments in preparation for another scan line. It is up to the printer control circuits to break down an image into individual scan lines, then direct the writing mechanism accordingly.

Lasers

Lasers have been around since the early 1960s, and have developed to the point where they can be manufactured in a great variety of shapes, sizes, and power output levels. To understand why lasers make such a useful writing mechanism, you must understand the difference between laser light, and ordinary white light as shown in Fig. 5-11.

Ordinary white light is actually not white. The light you see is composed of many different wavelengths, each traveling in its own directions. When these various wavelengths combine, they do so virtually at random. This makes everyday light very difficult to direct and almost impossible to control as a fine beam. As an example, take a flashlight and direct it at a far wall. You will see just how much white light can scatter and disperse over a relatively small distance.

The nature of laser light, however, is much different. A laser beam contains only one major wavelength of light (it is *monochromatic*). Each ray travels in the same direction and combines in an additive fashion (known as *coherence*). These characteristics make laser light easy to direct at a target as a hair-thin beam, with almost no scatter (*divergence*). Older EP printers used helium-neon (HeNe) gas lasers, but strong semiconductor laser diodes have essentially replaced gas lasers in almost all laser printing applications.

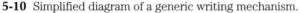

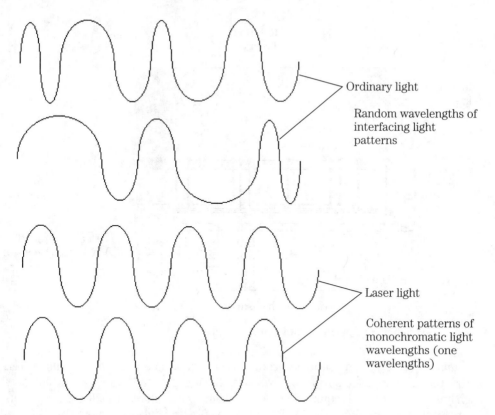

Ordinary light

Random wavelengths of interfacing light patterns

Laser light

Coherent patterns of monochromatic light wavelengths (one wavelengths)

5-11 Ordinary light versus laser light.

Laser diodes are very similar to ordinary light-emitting diodes as in Fig. 5-12. When the appropriate amount of voltage and current is applied to a laser diode, photons of light will be liberated that have the characteristics of laser light (coherent, monochromatic, and low divergence). A small lens window (or *aperture*) allows light to escape, and helps to focus the beam. Laser diodes are not very efficient devices: they require a lot of power to generate a much smaller amount of

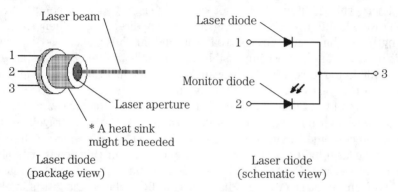

Laser beam

1
2
3

Laser aperture

* A heat sink might be needed

Laser diode (package view)

Laser diode

1

Monitor diode

2

3

Laser diode (schematic view)

5-12 Package and schematic diagram of a typical laser diode.

light power. This trade off is usually worthwhile for the small size, light weight, and high reliability of a semiconductor laser.

Generating a laser beam is only the beginning. The beam must be modulated (turned on and off) while being swept across the drum surface. Beam modulation can be accomplished by turning the laser on and off as needed (usually done with semiconductor laser diodes) as shown in Fig. 5-13, or by interrupting a continuous beam with an electro-optical switch (typically used with gas lasers that are difficult to switch on and off rapidly). Mirrors are used to alter the direction of the laser beam, and lenses are used to focus the beam and maintain a low divergence at all points along the beam path. Figure 5-13 is one illustration of a laser writing mechanism, but it shows the complexity that is involved. The weight of glass lenses, mirrors, and their shock mountings have kept EP laser printers bulky and expensive.

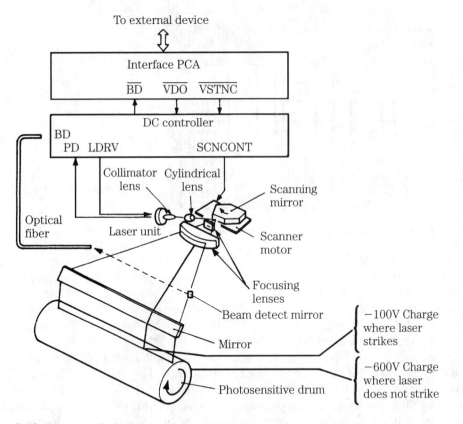

5-13 Diagram of a laser/scanning system. Hewlett-Packard Co.

Alignment has always been an unavoidable problem in complex optical systems such as Fig. 5-13. Consider what might happen to the beam if any optical component should become damaged or fall out of alignment—focus and direction problems could render a drum image unintelligible. Realignment of optical systems is virtually impossible without special alignment tools, and is beyond the scope of this book. Finally, printing speed is limited by the speed of moving parts, and the rate at which the laser beam can be modulated and moved.

LEDs

Fortunately, a photosensitive drum is receptive to light from many different sources. Even light from light-emitting diodes (LEDs) can expose the drum. By fabricating a series of microscopic LEDs into a single scan line as shown in Fig. 5-14, an LED can be provided for every possible dot in a scan line. For example, the ROHM JE3008SS02 is an LED print bar containing 2,560 microscopic LEDs over 8.53 inches. This number of LEDs equates to 300 dots per inch. Each LED is just 50×65 micrometers (μm), and they are spaced 84.6 μm apart.

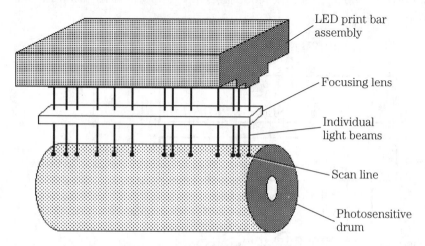

LED print bar
assembly

Focusing lens

Individual
light beams

Scan line

Photosensitive
drum

5-14 Diagram of an LED print bar in operation.

The operation of an LED print bar such as the one shown in Fig. 5-15 is remarkably straightforward. An entire series of data bits corresponding to each possible dot in a horizontal line is shifted into internal digital circuitry within the print bar. Dots that will be visible are represented by a logic 1, and dots that are not visible will remain at logic 0. For a device such as the JE3008SS02, 2,560 bits must be entered for each scan line.

After a complete line of data has been loaded through the *DIN* (data in) pin, the LEDs must be fired. This is performed in segments to reduce the power surges that would be generated if every LED were fired together. The JE3008SS02 is divided into four segments of 640 dots. A trigger signal (or *strobe*) can be applied to STR1 through STR4. The signal passes data to driver circuits of each segment. LEDs that illuminate will discharge latent points on the drum surface. LEDs that do not light will have no effect. Each strobe is fired sequentially until all four segments have been strobed. All 2,560 dots can be scanned in under 2.5 milliseconds (ms). The drum is incremented ¹⁄₃₀₀ of an inch, and a new scan line can be loaded into the print bar.

You can probably see the advantages of an LED print bar system over a laser approach. There are no moving parts involved in light delivery—no mirror motor to jam or wear out. The printer can operate at much higher speeds because it does not have to overcome the dynamic limitations of moving parts. There is only one focusing lens

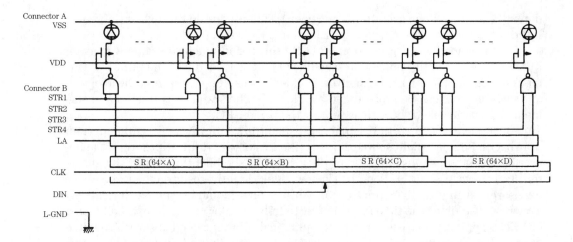

5-15 Partial schematic of an LED print bar. ROHM Corporation.

between the print bar and drum. This greatly simplifies the optics assembly, and removes substantial weight and bulk from the printer. An LED system overcomes almost all alignment problems, so a defective assembly can be replaced or aligned quickly and easily.

LCSs

A print bar does not necessarily have to generate its own light. Liquid crystal shutters (LCSs) control the transfer of light from a single fluorescent lamp source to the drum surface as shown in Fig. 5-16. Instead of LEDs, an array of individual liquid crystal shutters is fabricated into a single scan line—one shutter for every possible

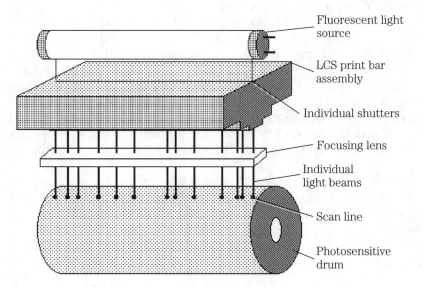

5-16 Diagram of an LCS print bar in operation.

dot in a scan line. When a shutter is on (open), light is allowed through at that point. If the shutter is off (closed) light at that point is blocked off.

Data is loaded and triggered in an LCS print bar in much the same way as for LED systems. An entire horizontal scan line of data is loaded into the print bar. One bit of data is supplied for each shutter. Visible dots can be represented by a logic 1, and invisible dots can be represented by a logic 0. The scan line is triggered in segments by a series of strobe signals sent by the printer ECP.

At the time of this writing, liquid crystal shutters suffer from several important drawbacks. First, the light source is very critical to proper drum exposure. If the fluorescent light source becomes old or dirty, it might shed light unevenly, so light intensity can vary along its length. Uneven light directly results in uneven drum exposure—even if the LCS print bar is working fine. Liquid crystal has a relatively slow response time (the time for a shutter to fully open or close) compared to laser or LED scanning. As a result, maximum practical printing speed is severely limited to just a few pages per minute. Finally, resolution also is limited. Current liquid crystal fabrication techniques allow no more than 300 dots per inch. Unless these limitations are overcome, LCS technology will never reach the acceptance that LED print bars have enjoyed.

The electrophotographic cartridge

Electrostatic printers mandate the use of extremely tight manufacturing tolerances to ensure precise, consistent operation. A defect of even a few thousandths of an inch could cause unacceptable image formation. Even the effects of normal mechanical wear can have an adverse effect on print quality. Many key IFS components would have to be replaced every 5,000 to 10,000 pages to maintain acceptable performance. Clearly it would be undesirable to send your printer away for a complete (and time-consuming) overhaul every 10,000 pages.

To ease manufacturing difficulties and provide fast, affordable maintenance to every ES printer user, critical components of the IFS, as well as a supply of toner, are assembled into a replaceable electrophotographic (or EP) cartridge. As Fig. 5-17 shows, a typical EP cartridge contains the toner roller, toner supply, debris cavity, primary corona (and primary grid), photosensitive drum, and cleaning blade assembly. All necessary electrical connectors and drive gears are included. By assembling sensitive components into a single replaceable cartridge, printer reliability is substantially improved by preventing problems before they become noticeable. The cost of an EP cartridge is low enough to consider it a disposable part.

A typical EP cartridge can produce 200 to 5,000 printed pages. The exact number varies depending upon just how much toner is available, and which critical parts are placed in the cartridge—highly integrated EP cartridges will last longer than simple cartridges. Because toner is comprised partially of organic materials, it has a limited useful life (often six months after the cartridge is removed from its sealed container).

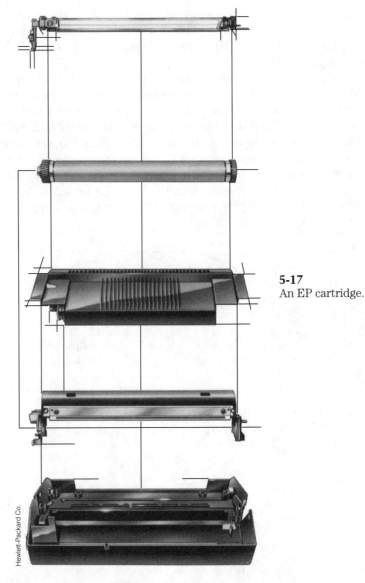

Hewlett-Packard Co.

5-17
An EP cartridge.

Protecting an EP cartridge

As you might imagine, the precision components in an EP cartridge are sensitive and delicate. The photosensitive drum and toner supply are particularly sensitive to light and environmental conditions, so follow the handling and storage guidelines.

The photosensitive drum is coated with an organic material that is extremely sensitive to light. Although a metal shroud covers the drum when the cartridge is exposed, light might still penetrate the shroud and cause exposure (also known as *fogging*). Deactivating the printer for a time will often eliminate mild fogging. Do not remove or interfere with the proper action of the shroud in open light unless absolutely necessary, and then only for short periods. Defeating the shroud will certainly fog the drum. You might need to place a seriously fogged cartridge in a dark area for several days. Never expose the drum to direct sunlight—direct sunlight can permanently damage the drum coating.

Avoid extremes of temperature and humidity. Temperatures exceeding 40°C can permanently damage an EP cartridge. Extreme humidity is nearly as dangerous. Do not allow the cartridge to become exposed to ammonia vapors or other organic solvent vapors— they break down the drum coating very quickly. Finally, keep a cartridge secure and level. Never allow it to be dropped or abused in any way.

As the toner supply diminishes, it might be necessary to redistribute remaining toner so that it reaches the toner roller. Because toner is available along the entire cartridge, it must be redistributed by rocking the cartridge back and forth along its long axis as shown in Fig. 5-18. If you tip a cartridge upright, remaining toner will fall to one end and cause uneven distribution.

Electrophotographic (EP)
cartridge assembly

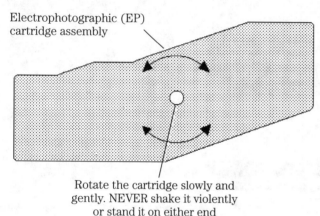

5-18
Redistributing toner in an EP
cartridge.

Rotate the cartridge slowly and
gently. NEVER shake it violently
or stand it on either end

6
CHAPTER

Power supplies

All electrical and electronic components in your laser printer (Fig. 6-1), as well as every other piece of electronic equipment, require electrical energy to function. Energy is always supplied in the form of voltage and current, and there must be adequate amounts of both to ensure the proper operation of each component. Unfortunately, ac power available in your home, shop, or office, is not directly compatible with the components in your printer. As a result, line power must be manipulated and converted into values of voltage and current that are suitable for your specific equipment. Conversion is the task of a *power supply*. You will find three types of power supplies in your laser printer: a dc supply, an ac supply, and a high-voltage supply.

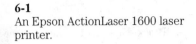

6-1
An Epson ActionLaser 1600 laser printer.

The name *power supply* is misleading. A *PS* (power supply) does not actually *create* power. Instead, it converts commercially generated ac power into one or more voltage levels better suited to particular tasks and components. Power supplies are generally rugged and reliable devices—so much so that they are often over-looked or disregarded as possible problem sources. Luckily, most supplies are reasonably simple to follow, and can be repaired or replaced with ease. **Before you attempt a power-supply repair, be sure to review the hazards of high-voltage electricity outlined in chapter 4.**

Many different PS circuits exist. Each is designed to best suit the needs of the specific circuits that they must supply. In spite of the array of PS arrangements, there are only two typical operating modes—*linear* and *switching*. Both modes refer to the way a supply controls its output(s). This chapter shows you the construction, operation, characteristics, and repair considerations for both power supply types. Background for high-voltage supplies is not discussed here because of the dangers and stringent component requirements associated with high-voltage systems, but troubleshooting information for the supplies is presented.

Power supplies—ac and linear dc

The term *linear* means line or straight. As shown in the block diagram of Fig. 6-2, a linear PS operates in essentially a straight line from ac input to dc output. Components and power capacity can vary between manufacturers and models, but all linear dc power supplies will contain the same three basic subsections: a *transformer*, *rectifier*, and *filter*. The *regulator* block also is found in the majority of linear power supplies, but it is not mandatory for a minimum working supply. Alternating-current power supplies are generally simple transformers to provide one or more levels of ac to heat the fusing assembly.

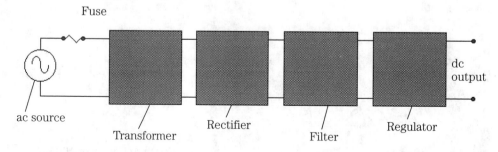

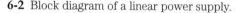

6-2 Block diagram of a linear power supply.

Transformers

A *transformer* is used to alter the ac voltage and current characteristics of ac input power, allowing ac to be converted into more useful levels of voltage and current. This important process of transformation is accomplished through the principles of *magnetic coupling* as shown in the schematic of Fig. 6-3.

Transformers use two coils of solid wire wrapped along opposite sides of a common metal structure (called a *core*). Although Fig. 6-3 only shows two leads for each coil, many transformers offer available leads (or *taps*) from both the primary and secondary coils. The core is often built from laminated plates of *permeable* material

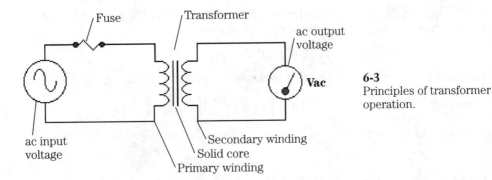

6-3
Principles of transformer operation.

(metal that can be magnetized). The core serves not only as a physical base, but it is critical in concentrating magnetic fields around the transformer as well.

An ac voltage (or *primary voltage*), usually 120 Vac, is applied across the primary winding of a transformer. Alternating current voltage will cause current to fluctuate through the primary winding. In turn, this sets up a varying magnetic force field in the primary. The core concentrates this magnetic field and helps to transfer magnetic force to the secondary winding. Note that a solid core is not mandatory—magnetic coupling between two coils can occur across an air gap—but solid cores make coupling much more efficient. This fluctuating magnetic field in the core cuts across the secondary winding, where it induces a secondary ac voltage between its terminals. Figure 6-3 shows an ac voltmeter measuring this voltage and shows the transformer principle.

Voltage across the secondary winding is directly proportional to the ratio of primary-to-secondary windings. For example, if there are 1,000 turns of wire in the primary coil and 100 turns of wire in the secondary coil, the ratio (called *turns ratio*) would be 10:1. Because there are fewer secondary windings than primary windings, that transformer will be known as a *step-down* transformer. If 120 Vac were applied across the primary, its secondary would ideally yield

$$12 \text{ Vac} \left[\frac{120 \text{ Vac}}{\left(\dfrac{10}{1}\right)} \right]$$

If the ratio were reversed with 100 primary turns and 1,000 secondary turns, the transformer would be a 1:10 *step-up* device. An input of 12 Vac to this primary would result in an output of

$$120 \text{ Vac} \left[\frac{12 \text{ Vac}}{\left(\dfrac{1}{10}\right)} \right]$$

across the secondary. An *isolation transformer* has a 1:1 turns ratio—the number of primary and secondary turns is equal, so secondary voltage will ideally equal primary voltage.

A transformer also steps current, but current is stepped in reverse of the voltage ratio. If voltage is stepped up, current is stepped down by that same ratio, and vice versa. In this way, power taken from a transformer secondary will roughly equal the power provided to its primary.

As an example, suppose the transformer of Fig. 6-3 has 120 Vac at 0.1 A supplied to its primary. Primary power would then be ($P = I \times V$) or

$$12 \text{ W } (120 \text{ Vac} \times 0.1 \text{ A}).$$

With a 10:1 step-down transformer, secondary voltage would be 12 Vac, but

$$1 \text{ A} \left[0.1 \times \left(\frac{10}{1} \right) \right]$$

of current would be available. This results in a secondary power of

$$12 \text{ W } (12 \text{ Vac} \times 1 \text{ A}).$$

On paper, power output always equals the power input. In reality, however, output power is always slightly less than input power due to losses in the core and coil resistance. Severe losses can cause excessive heating in the transformer. The ratio of output power to input power (P_o/P_i) is known as *efficiency*. Most solid-core transformers can reach 80 to 95% efficiency—but never 100%.

You might wonder why transformers will not step dc voltages. After all, dc can produce a strong magnetic field in solenoids (impact print wires for example). Although dc can produce a field, a magnetic field must fluctuate as a function of time to induce a *potential* (voltage) on another conductor. Direct current would certainly magnetize the primary winding, but without constant fluctuation, no voltage would be induced across the secondary winding.

The ac power supplies

Now that you understand how a transformer works, you have the essential knowledge to understand the concepts of an ac power supply that is little more than the output from a transformer. As you saw in the last chapter, heat is needed to fuse toner to the paper. Heat is typically provided by a quartz lamp inserted into the upper of two fusing rollers. Heat requires substantial power. Transformers are efficient devices capable of handling large amounts of power, so the heating lamps are almost always powered by ac provided from a transformer. Keep in mind that ac might be supplied by a stand-alone transformer, or taken from a tap off the transformer powering your printers dc supply. Because ac sources and dc supplies are so often integrated into the same assemblies, symptoms and solutions to ac supply problems will be covered with dc linear supply problems.

Rectifiers

Voltage across the transformer secondary is still in an ac form—that is, the polarity swings between positive and negative voltages. Alternating current must be converted into dc before it can be used by most electronic components. This conversion is known as *rectification*, where only one polarity of the input is allowed to reach the output. Although a rectifier output varies greatly, the polarity of its signal will always remain within one polarity—thus the term *pulsating dc*. Diodes are ideal for use in rectifier circuits because they only allow current to flow in one direction. You will encounter three classical types of rectifier circuits: half wave, full wave, and diode bridge.

A *half-wave* rectifier circuit is shown in Fig. 6-4. It is the simplest and most straightforward type of rectifier circuit because it only requires one diode. As secondary ac voltage exceeds the turn-on voltage of the diode (about 0.6 V for a silicon diode), it begins to conduct current. The current generates an output that mimics the positive half of the ac input. If the diode were reversed, its output would be reversed. The disadvantage of this type of rectifier is that it is inefficient—it only deals with half of its ac input—the other half is basically ignored. The resulting gap between pulses results in a lower average output and a higher amount of *ripple* (ac noise) contained in the final dc signal. Half-wave rectifiers are rarely used in modern power supplies.

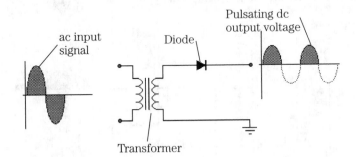

6-4 A half-wave rectifier circuit.

Full-wave rectifiers such as the one shown in Fig. 6-5 offer some substantial performance advantages over the half-wave design. By using two diodes in the configuration shown, both polarities of the ac secondary voltage input can be rectified into pulsating dc. Because a diode is at each terminal of the secondary, polarities at each diode will be opposite as shown. When the ac signal is positive, the upper diode conducts, but the lower diode is cut off. When the ac signal becomes negative, the lower diode conducts, but the upper diode is cut off. One diode is always conducting, so there are no gaps in the final output signal. Ripple levels are lower and the average dc output voltage is higher. The disadvantage to a full-wave rectifier is its transformer requirement. A center-tapped secondary is needed to provide a ground reference for the supply, which often takes a larger transformer. Large transformers are not popular with printer designers trying to reduce weight and bulk.

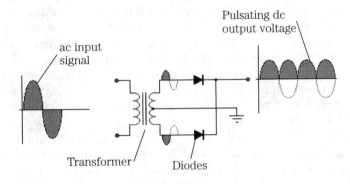

6-5 A full-wave rectifier circuit.

Diode bridge rectifiers use four rectifier diodes to provide full-wave rectification without the troubles of a center-tapped transformer. Figure 6-6 shows a typical bridge rectifier stage. Alternating current from the transformer secondary is connected to a series of diodes arranged in a *Wheatstone bridge* fashion. Diodes D1 and D2 provide the forward current paths, and D3 and D4 offer isolation between secondary voltage and a common reference point that serves as ground. When ac voltage is positive, diode D1 conducts because it is forward biased, and D4 provides isolation versus ground. As ac voltage becomes negative, D2 conducts while D3 supplies isolation versus ground. The complete bridge generates a full-wave pulsating dc output. Bridges are by far the most popular type of rectifier circuit.

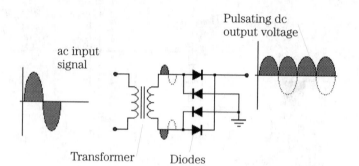

6-6 A bridge rectifier circuit.

Filters

By strict technical definition, pulsating dc is dc because voltage polarity remains consistent (even if its magnitude does change). Unfortunately, even pulsating dc is unsuitable for any type of electronics power source. Voltage levels must be constant over time to operate electronic devices properly. A *filter* is used to achieve a smoothed dc voltage as shown in Fig. 6-7.

Capacitors are typically used as filter elements because they act as voltage storage devices—almost like light-duty batteries. When pulsating dc is applied to a capacitive filter as in Fig. 6-7, the capacitor charges with current supplied from the rectifier. Ultimately, the capacitor charges to the peak value of pulsating dc. When a pulse falls off its peak (back toward zero), the capacitor will continue to supply current to a load. This action tends to hold up the output voltage over time—dc is filtered.

However, filtering is not a perfect process. As current is drained away from the capacitor by its load, voltage across the filter also will drop. Voltage continues to drop until a new pulse of dc recharges the filter for another cycle. This repetitive charge and discharge results in regular fluctuations of the filter output. These fluctuations are known as *ripple*. Ripple is an undesirable component of a smoothed dc output.

Figure 6-7 also shows a graph of voltage versus time for a typical filter circuit. The ideal dc output would simply be a constant, flat line at all points in time. In reality, there will always be some amount of filter ripple. Just how much ripple depends upon the load. For a light load (a high resistance that draws relatively little current), discharge is less between pulses, so ripple also is lower. A large load (a low resistance

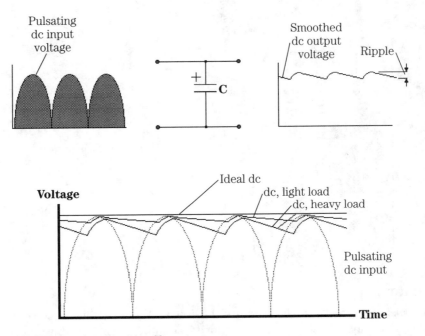

6-7 Effects of a capacitive filter stage.

that draws substantial current) requires greater current, so discharge (and ripple) is greater between pulses. The relationship of dc pulses is shown for reference.

Additional filtering can be accomplished by adding more capacitance to the filter stage. More capacitance holds more charge, so load can be supplied with less discharge. As a general rule, more capacitance results in less ripple and vice versa. Although this is true in theory, there are some practical limits to just how much capacitance can be used in a power filter. Size is always a big concern. Capacitors larger than 4,700 μF are large and cumbersome. Above 10,000 μF, a filter can accept so much charging current on its initial charge (known as *inrush current*) that it might seem like a short circuit. Excessive inrush current can blow a fuse or even damage the rectifier stage.

Before attempting to work on a power supply, you must understand the potential for a filter shock hazard. Power capacitors accumulate a substantial amount of electrical charge and hold it for a long time. If you touch the leads of a charged capacitor, current will flow through your body. Although this is almost never dangerous, it can be very uncomfortable or result in a slight burn.

To remove any stored charge in your filter stage, the charge must be bled away in a controlled fashion as shown in Fig. 6-8. A large-value resistor (called a *bleeder resistor*) can be connected across the filter. The resistor will slowly drain away any remaining charge. Note that some filter capacitors might already be built with a bleeder resistor. If a load remains connected across the filter, that also will discharge the filter after power is removed. Never attempt to discharge a capacitor using a screwdriver or wire. The sudden release of energy can actually weld a wire or screwdriver blade right to the capacitor terminals, as well as damage the capacitor internal structure.

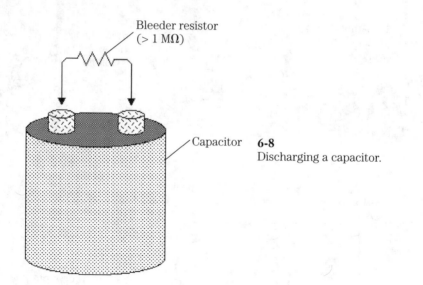

6-8
Discharging a capacitor.

Regulators

A transformer, rectifier, and filter are the essentials of every linear power supply. These parts combined will successfully convert ac into dc that can drive most basic electrical and electronic components. There are several troubles with these simple *unregulated* supplies that make them undesirable. First, ripple is always present at a filter output. Under some circumstances, this can cause erratic operation in even the most forgiving ICs (integrated circuits). Second, output voltage varies with load. While load is fairly light, this effect might be negligible, but the effects of heavy loads also can cause unpredictable circuit performance. The filter output must be stabilized to eliminate effects of ripple and loading. Stabilizing the output is the task of a *regulator*.

Linear regulation is as its name implies—current flows from the regulator input to the outputs as shown in Fig. 6-9. Voltage that is supplied to the regulator input must be somewhat higher than the desired output voltage (usually by several volts). Internal circuitry within the regulator manipulates input voltage to produce a steady, consistent output level over a fairly wide range of loads and input voltages. If input voltage drops below some minimum value, the regulator falls out of regulation. In that case, the output signal follows the input signal—including ripple.

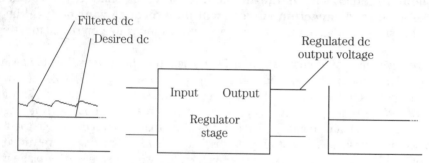

6-9 Block diagram of a generic regulator.

To maintain a constant output voltage, the linear regulating circuit (or IC) will throw away extra energy provided by the filter in the form of heat. To carry away the energy, most regulators are often attached to large metal heat sinks. *Heat sinks* carry heat to the surrounding air. Although linear regulation provides a simple and reliable method of operation, it also is very wasteful and inefficient. Typical linear regulators are only up to 50% efficient. For every 10 W of power provided to the supply, only 5 W is provided to the load. Much of this waste occurs in the regulation process. Switching regulation is much more efficient, but that subject is covered later in this chapter.

You might encounter many various types of regulator circuits. Figure 6-10 shows a very simple series voltage regulator constructed with discrete parts. Input voltage is applied to the zener diode (Zd) through a current-limiting resistor (R_z). The zener diode *clamps* voltage to its zener level. In turn, this zener potential turns on the power transistor that allows load current to flow. Output voltage equals zener voltage minus a small voltage drop (usually 0.5 to 0.7 V) from the transistor base-emitter junction. You can set the output voltage by changing the zener diode.

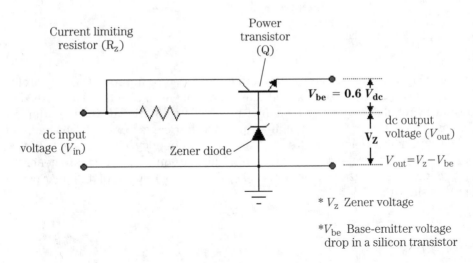

6-10 Diagram of a discrete series voltage regulator.

For the example of Fig. 6-10, suppose that input voltage is 10 V and you are using a 5.6 V zener diode. When power is applied to the circuit, zener voltage will be clamped at 5.6 V. Because input voltage is 10 V, the difference of 4.4 V (10 V – 5.6 V) will appear across the current-limiting resistor Rz. Zener voltage saturates the transistor, so its output will be 5.6 V minus the transistor base-emitter drop of 0.6 V, or 5.0 Vdc. As long as input voltage remains above the zener voltage, output voltage should remain steady regardless of load—output should be regulated. Load current can be substantial, so you will often find a power transistor used as the regulating transistor.

Regulator circuits can easily be fabricated as integrated circuits as shown in Fig. 6-11. Additional performance features such as automatic current limiting and

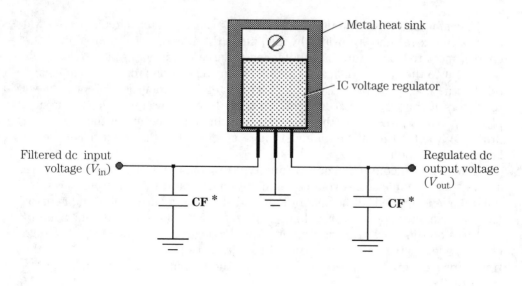

Metal heat sink

IC voltage regulator

Filtered dc input voltage (V_{in})

Regulated dc output voltage (V_{out})

CF *

CF *

* High-frequency filter capacitors

6-11 An IC voltage regulator.

over-temperature shutdown circuitry can be included to improve regulator reliability. Input voltage must still exceed some minimum level to achieve a steady output, but IC regulators are much simpler to use. One additional consideration for IC regulators is the addition of high-frequency filter capacitors at both the input and output. The capacitors filter out any high-frequency noise or signals that could interfere with regulator operation. *HF* (high-frequency) filters are generally small-value, nonpolarized capacitors (0.01 μF or 0.1 μF). A complete linear power supply is shown in Fig. 6-12.

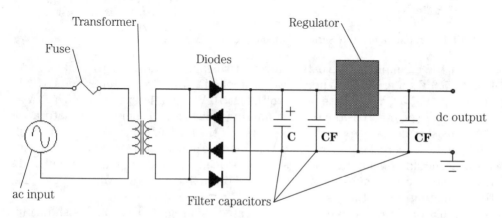

6-12 Schematic of a basic linear dc power supply.

Troubleshooting ac and linear dc supplies

Under most circumstances, linear power supplies are reasonably simple and straightforward to troubleshoot. A sound procedure is to use your voltmeter (or oscilloscope) to trace the presence of voltage through the supply. If an output has failed, start your measurements at the output and work back until you find the appropriate signal again. The following symptoms and troubleshooting procedures present more details. **If you should determine a power supply to be defective, remember that it is always acceptable to replace the supply outright.**

Symptom 1 *Power supply is completely dead. Laser printer does not operate and no power indicators are lit.* Before you begin to disassemble the printer, check to make sure that you are receiving an appropriate amount of ac line voltage into the power supply. Use your multimeter to measure ac voltage at the wall outlet powering your printer. Normally, you should read between 105 and 130 Vac (210 to 240 Vac in Europe) for a power supply to operate properly. More or less ac line voltage can cause the power supply to malfunction. **Exercise extreme caution whenever measuring ac line voltage levels. Review the hazards of electricity as discussed in chapter 4.**

When you determine that an appropriate amount of ac is available at the printer, the fault probably exists at the printer. Check the printer power switch to be sure that it is turned on. Even though it sounds silly, this really IS a common oversight. Next, check the printer main line fuse that is often located closely to the ac line cord connector. Unplug the printer *before* removing the fuse for examination. You should find the fusible link intact, but it is not always possible to see the entire link. Use your multimeter to measure continuity across the fuse. Normally, a working fuse should read as a short circuit (0 Ω). If you read infinite resistance, the fuse is defective and should be replaced. **Use caution when replacing fuses. Use *only* fuses of the same rating.** If a new fuse fails immediately when replaced, it suggests a serious failure (such as a short circuit) elsewhere in the power supply or printer. Do not continue to replace fuses if they continually fail.

If ac voltage and the fuse is intact, you must disassemble your printer and work on the power supply. **Take all precautions to protect yourself from ac and high-voltage hazards.** Check all connectors and wiring leading to or from the supply to rule out a broken wire or loose connector.

Turn on printer power and use your multimeter to measure dc output voltage(s) from the supply. Most printed circuit markings will give you some indication of what voltage should exist at each respective output. A low or nonexistent output indicates a problem. Make sure that the output is not being shorted by its load. Disconnect the supply from its load and measure its output(s) again. If your readings climb up to a normal level, there might be a short circuit somewhere in the printer electronics. If readings remain low, you will have to troubleshoot the supply. If you do not have the inclination or skill to test the supply in more detail, replace the entire supply assembly.

For this procedure, refer to the diagram of Fig. 6-12. If the voltage output is completely zero, check for the presence of a dc (low voltage) protection fuse in the output circuit. Some might be normally sized fuses in the power supply, but other fuses might be installed in the printer electronic control package (ECP). If supply outputs measure okay but the printer still does not function, look carefully for any subminiature or *pico* fuses (resembling carbon film resistors) that might be defective.

Of all the components in your supply, the regulator has the greatest stress. Use your multimeter to measure the dc input to the regulator. You should read several volts greater than the expected output. For example, a regulator with an output of +5 Vdc requires an input of +7 or +8 Vdc. When the regulator input voltage is correct, but its output is not, the regulator is probably defective and should be replaced. A low or nonexistent regulator input suggests a faulty filter or rectifier.

A shorted power filter capacitor can pull down the output from a rectifier. Unplug the printer, remove at least one capacitor lead from its circuit, and test the capacitor as discussed in the test equipment section of chapter 3. Replace any filter capacitor that appears open or shorted. Any power capacitor that appears hot or smells strange is a clear indication of trouble.

Inspect the rectifier circuit carefully. A faulty rectifier diode can completely disable your supply. Unplug the printer and test each rectifier diode as discussed in the test equipment section of chapter 3. When a bridge rectifier fails, you will usually find the two forward diodes open circuited. Replace any diodes that appear open or short circuited. If your rectifier is built into a potted bridge module, the entire module must be replaced.

Finally, turn on printer power and check the ac voltages at the transformer primary and secondary windings. **Use caution when measuring ac.** You should find about 120 Vac across the primary and some lower amount of ac (usually between 8 and 30 Vac) across the secondary. An open circuit in either winding can prevent any secondary output. Be careful to check for shorted transformer windings. Be suspicious of a transformer that becomes very hot after a short period of use, or one that emits an audible 60 Hz hum. Such a transformer might be developing a short circuit.

Consider the possibility of a PC board failure, especially if the laser printer failed after being dropped or abused. Faulty soldering at the factory (or on a previous repair bench) also can cause a PC board problem. As Fig. 6-13 shows, there are three different kinds of problems that can plague a printed circuit: lead pull-through, trace break, and board crack.

Lead pull-through occurs when a component lead or wire is ripped away from its through hole. Often, the soldering at the printed circuit pad might appear perfectly normal, but there will be a hole in the middle where the lead was. The lead also might remain within its printed circuit hole, but not be fully connected. This kind of problem can easily result in bizarre, intermittent behavior, but it can be repaired simply by reheating the solder joint to reconnect the lead.

Trace break occurs commonly around large or awkward components that are too well anchored to the PC board to experience lead pull-through. Instead, physical force will break the solder pad away from its trace. Trace breaks are difficult because they are usually so fine and clean that you might not see them upon a visual inspection. You have to spot them by wiggling each lead individually. When a solder pad moves, but its trace does not, you will see the break location. Jumper between two adjacent solder pads to reliably repair this failure. Do not attempt to solder or jumper across the break itself. Chemical coatings applied to printed circuit boards prevent solder from sticking to trace areas.

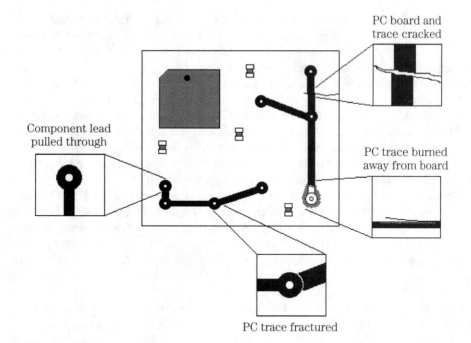

6-13 Typical failures in PC boards.

Board cracks accompany such physical traumas as drops or other abuse. Impact forces can actually crack the circuit board, which can split any traces that might run across the crack. Luckily, board cracks are relatively easy to spot. As with trace breaks, you must jumper between two adjacent solder pads to make a reliable repair.

Symptom 2 *Supply operation is intermittent. Printer operation cuts in and out along with the supply.* Begin by inspecting the ac line voltage powering your printer. If your line cord is loose at the wall plug or printer, it can play havoc with printer operation. **Use extreme caution when dealing with ac line voltages to prevent injuries from electrical shock.** Check the integrity of any ac connections attached to the printer power supply. Also check the dc output connectors attached to the printer internal circuitry. Tighten or replace any connectors that appear to be loose.

Check the power supply PC board for any signs of failure—especially if the printer has begun malfunctioning after a drop or other physical abuse. Faulty soldering connections from the factory (or from your workbench) also can cause printed circuit failures. Review Fig. 6-13 for three typical printed circuit problems.

Lead pull-through is a fault that occurs anywhere a component lead or wire is soldered into a through hole. Sudden, sharp force applied to the lead can overcome connection strength and rip the lead right away from its solder joint. Ripping out a lead might or might not pull the lead from its hole entirely. Trace breaks can happen anywhere a round solder pad meets a printed trace. Sudden impacts that do not cause lead pull-through might cause a hairline fracture between the solder pad and its printed trace. Hairline fractures can be particularly difficult problems because you might not be able to see trace breaks on visual inspection. You might have to wiggle each solder pad gently to reveal any trace breaks. Board cracks are fairly obvious problems, but large cracks can sever many traces—it really depends on the crack size and board complexity. Usually, breaks or cracks can be corrected by soldering jumpers between the solder pads across each defect. Of course, you might decide to replace the supply outright rather than try to repair PC board problems.

Consider the possibility of thermal intermittents if your printer works fine when it is first turned on but fails after some period of operation. Often, the printer must then sit for a time with all power off before it can be used. Test for thermally intermittent components by spraying suspect parts with a liquid refrigerant (available from almost any electronics parts store or mail-order house).

Begin by exposing the power supply. Apply power and operate the printer until it fails. Use your multimeter to measure each supply output before AND after it fails, so you will know which outputs are failing. When you have identified a faulty output, check its regulator for excessive heat. Never touch live components that might be hot or carrying high voltages—it is a certain opportunity for injury! Instead, smell around the regulator for any trace of smoke or unusually heated air. Spray the regulator with refrigerant, wait a moment, and recheck your output voltage. If normal voltage returns temporarily, you have isolated the problem. Replace the faulty regulator. Keep in mind that you might have to spray a component several times to cool it properly.

Filter components and rectifier diodes are rarely subject to thermal problems. Transformer windings can open or short due to excessive heat, but only after a long period of breakdown. Use your multimeter to measure voltages through the remainder of the supply to track down any further problems. **If your tests are inconclusive, you might decide to replace the supply**.

Symptom 3 *Laser printer is not operating properly. It might be functioning erratically or not at all. Power indicators might or might not be lit.* Use your multimeter to measure the ac line voltage reaching the printer. Under normal circumstances, you should measure 105 to 130 Vac (some European countries use 210 to 240 Vac). On the average, 120 Vac should be available. If line voltage drops below 105 Vac, power supply outputs can begin to fall out of regulation. As a result, printer circuits might not receive enough voltage or current to ensure proper operation, which can cause erratic operation that can disable (or even damage) the printer. High input voltages (over 130 Vac) can force more current into the supply than desirable. Additional current generates heat that can cause premature breakdowns in the power supply.

Unplug the laser printer and check for any loose connectors or wiring that might be interrupting circuit operation. The connectors might have been improperly installed at the factory, or you might have re-installed them incorrectly during a previous repair effort. Turn on printer power and use your multimeter and measure voltage at each supply output. If you locate a defective output, troubleshoot the supply from its output, back to its transformer as discussed below. If all supply outputs appear correct, there might be a fault in the printer electronic control package, so troubleshoot your printer electronics according to the procedures outlined in chapter 9.

If you detect a faulty supply output, use your multimeter to measure input voltage at that regulator. It should be several volts higher than the expected regulator output. When a regulator input appears normal, but its output does not, try replacing the regulator. Low regulator input voltage (or no input voltage) might be caused by a fault in the filter or rectifier stages. Unplug the printer and check your filter capacitor(s) for open or short circuits as discussed in the test equipment section of chapter 3. Replace any filter capacitors that appear defective. If the filter capacitor(s) appear intact, check each rectifier diode. Replace any rectifier diodes that appear faulty. You also might replace the power supply outright.

Symptom 4 *Fusing quality is intermittent or poor (toner smudges easily), or fuser fails to reach operating temperature within 60 to 120 seconds.* Toner must be heated to about 180°C to bond to paper properly. Normally, voltage supplied from the ac power supply will allow the quartz lamp to heat. If ac is low or intermittent, the quartz lamp might not be able to reach or hold its operating temperature. Ideally, you should be able to run your finger tips across the surface of a printed page without smudging the page. If the print smudges, the toner is not being heated enough. There might be several reasons for this. The quartz lamp might be failing, the electronic temperature control or sensor might be faulty, or the ac supply might be low.

Unplug the printer, expose the ac supply area, and make sure that all connections are secure. Apply power to the printer and use your voltmeter to measure the ac voltage being provided to the quartz lamp. **Use extreme caution to avoid receiving an electric shock or a burn from the quartz lamp—remember that temperatures can approach 180°C.** The correct ac voltage level will usually be marked on the transformer output. If voltage is correct, the problem is probably in the lamp or temperature control (refer to chapter 7 for more details). If the ac voltage level is low or absent, measure the ac input that should be approximately 120 Vac (220 Vac in many European countries). If ac input is low, or if ac input is correct but ac output is low or absent, replace the ac power module or transformer.

If the ac supply is absent, the quartz lamp has failed, or the electronic temperature control or sensor has failed completely, the fuser will not even approach its operating temperature. This problem will usually result in an error code being displayed on the laser printer control panel during initialization. Chapter 7 deals with fuser problems in more detail.

Construction and operation of dc switching supplies

The great disadvantage of linear power supplies is their tremendous waste. At least half of all power provided to a linear supply is literally thrown away as heat—most of this waste occurs in a regulator. Ideally, if there were just enough energy supplied to the regulator to achieve a stable output voltage for any given load, regulator waste could be reduced almost entirely.

Instead of throwing away extra input energy, a switching power supply senses the output voltage provided to a load, then switches the ac primary (or secondary) voltage on or off as needed to maintain steady levels. A block diagram of a typical switching power supply is shown in Fig. 6-14. There are many configurations possi-

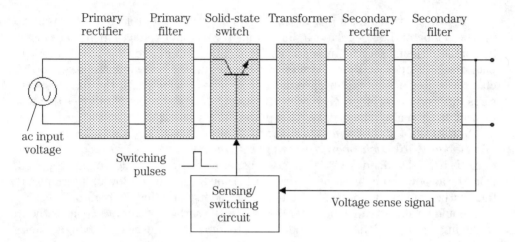

6-14 Block diagram of a switching power supply.

ble, but Fig. 6-14 shows one possible design. You can see the similarities and differences between a switching supply, and the linear supply shown in Fig. 6-2.

Alternating-current line voltage entering the supply is immediately converted to pulsating dc; then it is filtered to provide a primary dc voltage. Notice that ac is not transformed before rectification, so primary dc can reach levels approaching 170 V. Remember that ac is at least 120 V rms. Because capacitors charge to the peak voltage (peak = rms × 1.414), dc levels can be higher than your ac voltmeter readings. **This level of dc is as dangerous as ac line voltage, and should be treated with extreme caution.**

On start up, the switching transistor is turned on and off at a high frequency (usually 20 to 40 kHz), and a long duty cycle. The switching transistor breaks up this primary dc into chopped dc that can now be used as the primary signal for a step-down transformer. The duty cycle of chopped dc will effect the ac voltage level generated on the transformer secondary. A long duty cycle means a larger output voltage (for heavy loads) and a short duty cycle means lower output voltage (for light loads). *Duty cycle* itself refers to the amount of time that a signal is on compared to its overall cycle. Duty cycle is continuously adjusted by the sensing/switching circuit. You can use an oscilloscope to view switching and chopped dc signals.

Alternating-current voltage produced on the transformer secondary winding (typically a step-down transformer) is not a pure sine wave, but it alternates regularly enough to be treated as ac by the remainder of the supply. Secondary voltage is rectified and filtered again to form a secondary dc voltage that is actually applied to the load. Output voltage is sensed by the sensing/switching circuit that constantly adjusts the chopped dc duty cycle. Figure 6-15 shows a more practical circuit for the classical switching supply.

As load increases on the secondary circuit (more current is drawn by the load), output voltage tends to drop. This is perfectly normal—the same thing happens in every unregulated supply. However, a sensing circuit detects this voltage drop and increases the switching duty cycle. In turn, the duty cycle for chopped dc increases,

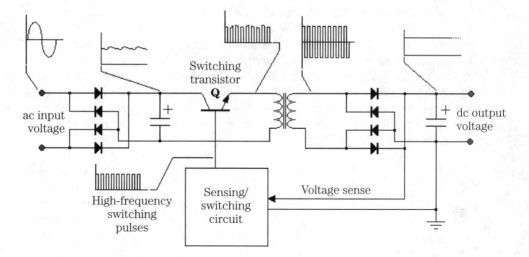

6-15 Schematic diagram of a basic switching power supply.

which increases the voltage produced by the secondary winding. Output voltage climbs back up again to its desired value—output voltage is regulated.

The reverse will happen as load decreases on the secondary circuit (less current is drawn by the load). A smaller load will tend to make output voltage climb. Again, the same actions happen in an unregulated supply. The sensing/switching circuit detects this increase in voltage and reduces the switching duty cycle. As a result, the duty cycle for chopped dc decreases, and transformer secondary voltage decreases. Output voltage drops back to its desired value—output voltage remains regulated.

Consider the advantages of a circuit such as Fig. 6-15. Current is only drawn in the primary circuit when its switching transistor is on, so very little power is wasted in the primary circuit. The secondary circuit will supply just enough power to keep load voltage constant (regulated), but very little power is wasted by the secondary rectifier, filter, or switching circuit. Switching power supplies can reach efficiencies higher than 85%, which is 35% more efficient than most comparable linear supplies. More efficiency means less heat is generated by the supply, so components can be smaller and packaged more tightly.

Unfortunately, there are several disadvantages to switching supplies that you must be aware of. First, switching supplies tend to act as radio transmitters. Their 20 to 40 kHz operating frequencies can interfere with radio and television reception and with circuits inside the printer. This potential problem is prevented in most switching supplies by covering or shielding with a metal casing. It is important that you replace any shielding removed during your repair. Strong *EMI* (electromagnetic interference) can disturb the printer. Second, the output voltage will always contain some high-frequency ripple. In many applications, this is not enough noise to present interference to the load. In fact, a great many printers use switching power supplies. Finally, a switching supply often contains more components and is more difficult to troubleshoot than a linear supply. This disadvantage is often outweighed by the smaller, lighter packaging of a switching supply, but replacement rather than repair is usually an economical decision.

Sensing and switching functions can be fabricated right onto an integrated circuit. IC-based switching circuits allow simple, inexpensive circuits to be built as shown in Fig. 6-16. Notice how similar this looks to a linear supply. Alternating-current line voltage is transformed (usually stepped down), then it is rectified and filtered before reaching a switch regulating IC. The IC chops dc voltage at a duty cycle that will provide adequate power to the load. Chopped dc from the switching regulator is filtered by the combination of choke and output filter capacitor to reform a steady dc signal at the output. The output voltage is sampled back at the IC that constantly adjusts the chopped dc duty cycle.

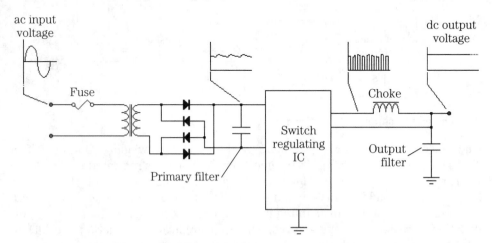

6-16 Schematic of a simple IC-based switching supply.

Troubleshooting dc switching supplies

Troubleshooting a switching power supply can be a complex and time-consuming task. Although the operation of rectifier and filter sections are reasonably straightforward, sensing/switching circuits can be complex oscillators that are difficult to follow without a schematic. Subassembly replacement of dc switching supplies are common. For this troubleshooting discussion, consider the IC-based switching supply of Fig. 6-17.

The STK7554 is a switching regulator IC manufactured as a 16-pin *SIP* (single in-line package). It offers a dual output of 24 Vdc and 5 Vdc. Notice that BOTH output waveforms from the STK7554 are 38 V square waves, but it is the *duty cycle* of those square waves that sets the desired output levels. The square wave amplitude simply provides energy to the filter circuits. Filters made from coils (or *chokes*) and high-value polarized capacitors smooth the square-wave input (actually a form of pulsating dc) into a steady source of dc. There will be some small amount of high-frequency ripple on each dc output. Smaller, nonpolarized capacitors on each output act to filter out high-frequency components of the dc output. Finally, note the resistor-capacitor-diode combinations on each output. The combinations form a surge and flyback protector that prevents energy stored in the choke from re-entering the IC and damaging it. Refer to Fig. 6-17 for the following symptoms.

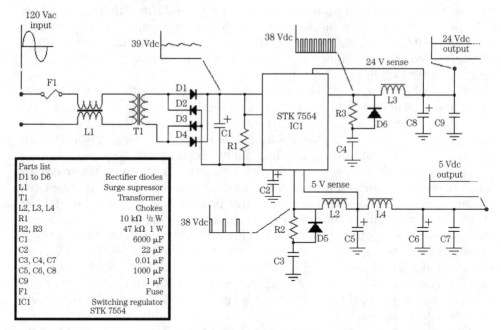

6-17 Schematic of a dual-output, IC-based switching DC power supply.

Symptom 1 *Power supply is completely dead. Laser printer does not oper-ate, and no power indicators are lit.* As with linear supplies, check the ac line volt-age entering the printer before beginning any major repair work. Use your multimeter to measure the ac line voltage available at the wall outlet powering your printer. **Use extreme caution whenever measuring ac line voltage levels. Review the hazards of electricity as discussed in chapter 4.** Normally, you should read be-tween 105 and 130 Vac to ensure proper supply operation. If you find either very high or low ac voltage, try the printer in an outlet that provides the correct amount of ac voltage. Unusual line voltage levels might damage your power supply, so pro-ceed cautiously.

If ac line voltage is normal, suspect the main power fuse at the printer. Most power fuses are accessible from the rear of the printer near the ac line cord, but some fuses might only be accessible by disassembling the printer. Unplug the printer and remove the fuse from its holder. You should find the fusible link intact, but use your multimeter to measure continuity across the fuse. A good fuse should measure as a short circuit (0 Ω), and a failed fuse will measure as an open circuit (infinity). Replace any failed fuse and re-test the printer. If the fuse continually fails, there is a serious defect elsewhere within the power supply or other printer circuits. If your printer has an ac selector switch that sets the supply for 120 Vac or 240 Vac opera-tion, be sure that switch is in the proper position.

Unplug the printer and disassemble it enough to expose the power supply clearly. Turn on the printer and measure each dc output with your multimeter or os-cilloscope. If each output measures correctly, then your trouble lies outside the sup-ply—perhaps in some connector or wiring that provides power to the printer electronics package. A low output voltage suggests a problem within the supply itself. Check each connector and all interconnecting wiring leading to or from the supply. Many switching supplies *must* be attached to a load for proper switching to occur. If the load circuit is disconnected from its supply, the voltage signal could shut down or oscillate wildly.

When supply outputs continue to measure incorrectly with all connectors and wiring intact, chances are that your problem is inside the supply. With a linear supply, you begin testing at the output, then work back toward the ac input. For a switching supply, you should begin testing at the ac input, then work toward the defective output. You could also replace the supply module outright.

Measure the primary ac voltage applied across the transformer. **Use extreme caution when measuring high-voltage ac.** The value should be approximately 120 Vac. If voltage has been interrupted in that primary circuit, you will read 0 Vac. Check the primary circuit for any fault that might interrupt power. Measure secondary ac voltage supplying the rectifier stage. It should read higher than the highest output voltage that you expect. For the example of Fig. 6-17, the highest expected dc output is 24 V, so ac secondary voltage should be several volts higher than this (for example, 28 Vac). If primary voltage reads correctly and secondary voltage does not, you might have an open circuit in the primary or secondary transformer winding. Replace the transformer or replace the entire supply.

Next, check the preswitched dc voltage supplying the switching IC. Use your multimeter or oscilloscope to measure this dc level. You should read approximately the peak value of whatever secondary ac voltage you just measured. For Fig. 6-17, a secondary voltage of 28 Vac should yield a dc voltage of about 38 Vdc (28 Vac rms × 1.414) Vdc. If this voltage is low or nonexistent, unplug the printer and check each rectifier diode, then inspect the filter capacitor. Component testing techniques are shown in chapter 3.

Use your oscilloscope to measure each chopped dc output signal. You should find a high-frequency square wave at each output (20 to 40 kHz) with an amplitude approximately equal to the preswitched dc level (38 to 39 V in this case). Set your oscilloscope to a time base of 5 or 10 μs/DIV and start your VOLTS/DIV setting at 10 VOLTS/DIV. Once you have established a clear trace, adjust the time base and vertical sensitivity to optimize the display.

If you do not read a chopped dc output from the switching IC, either the IC is defective, or one (or more) of the polarized output filter capacitors might be shorted. Unplug the printer and inspect each questionable filter capacitor as discussed in chapter 3. Replace any capacitors that appear shorted. Usually, filter capacitors fail more readily in switching supplies than in linear supplies because of high-frequency electrical stress, and the smaller physical size of most switching supply components. If all filter capacitors check out correctly, replace the switching IC or replace the entire supply. Use care when desoldering the old regulator. Install an IC socket (if possible) to prevent repeat soldering work, then just plug in the new IC.

Symptom 2 *Supply operation is intermittent. Laser printer operation cuts in and out with the supply.* Begin by inspecting the ac line voltage into your printer. Be sure that the ac line cord is secured properly at the wall outlet and printer. Make sure that the power fuse is installed securely. If the printer comes on at all, the fuse must be intact. Unplug the printer and expose your power supply. Inspect every connector or interconnecting wire leading into or out of the supply. A loose or improperly installed connector can play havoc with a printer operation. Pay particular attention to any output connections. Most often, a switching power supply must be connected to its load circuit to operate. Without a load, the supply might cut out or oscillate wildly.

Often, intermittent operation might be the result of a PC board problem such as the ones shown in Fig. 6-13. PC board problems are often the result of physical abuse or impact, but they also can be caused by accidental damage during a repair. Lead pull-through occurs when a wire or component lead is pulled away from its solder joint, usually through its hole in the PC board. This type of defect can easily be repaired by re-inserting the pulled lead and properly resoldering the defective joint. Trace breaks are hairline fractures between a solder pad and its printed trace. Such breaks can usually make a circuit inoperative, and they are almost impossible to spot without a careful visual inspection. Board cracks can sever any number of printed traces, but they are often very easy to spot. The best method for repairing trace breaks and board cracks is to solder jumper wires across the damage between two adjacent solder pads. You also might simply replace the power supply outright.

Some forms of intermittent failures are time or temperature related. If your printer works just fine when first turned on, but fails only after a period of use, then spontaneously returns to operation later on (or after it has been off for a while), you might be faced with a thermally intermittent component. A component might work when cool, but fail later on after reaching or exceeding its working temperature. After a printer quits, check for any unusually hot components. Never touch an operating circuit with your fingers—injury is almost certain. Instead, smell around the circuit for any trace of burning semiconductor or unusually heated air. If you detect an overheated component, spray it with a liquid refrigerant. Spray in short bursts for the best cooling. If normal operation returns, then you have isolated the defective component. Replace any components that behave intermittently or replace the entire supply. If operation does not return, test any other unusually warm components.

Symptom 3 *Laser printer is not operating properly. It might be functioning erratically or not at all. Power indicators might or might not be lit.* Use your multimeter and check the ac line voltage first. Normally, domestic U.S. laser printers require a line voltage between 105 to 130 Vac to ensure proper operation. If line voltage is low, the supply output(s) might not be able to maintain regulation. As voltage falls, circuitry in the printer might begin to behave erratically or not at all—it could even damage some delicate printer circuits.

Check all wiring and connectors leading to and from the power supply to be sure that everything is tight and installed correctly. Pay particular attention to connector orientation. Loose, missing, or incorrectly inserted connectors can easily disable your printer, or at least cause unpredictable operation. Switching power supplies usually require a load circuit to be connected. Otherwise, its output(s) might oscillate out of control or shut down totally.

Many switching power supplies are contained in a metal enclosure or shroud that is wired to chassis (earth) ground. The ground blocks (or *attenuates*) any electromagnetic interference (EMI) generated by the supply. Make sure that all original shielding is in place and securely wired to ground. If it is not, EMI might interfere with the operation of other printer circuits to cause erratic or random behavior.

Use your multimeter or oscilloscope to measure each supply output. If all outputs measure correctly, then your trouble is most likely in the printer electronics. Refer to the troubleshooting procedures for electronic circuits contained in chapter 9. Outputs that are low or nonexistent suggest a problem in the power supply itself. Trace the supply from ac input to dc output(s) or replace the supply.

Measure the ac voltage across your transformer primary winding. Under normal circumstances, it should read approximately 120 Vac. Check the secondary ac voltage from the transformer. You should measure an ac level that is higher than your largest expected dc output. For Fig. 6-17, the secondary voltage should be about 28 Vac. If the secondary voltage appears low or nonexistent, the transformer primary or secondary windings might be defective.

Measure the filtered dc voltage entering the switching circuit. You should read a voltage level approximately equal to the peak secondary voltage. In Fig. 6-17, your dc reading should be about 39 Vdc (28 Vac × 1.414). If this voltage is low or non-existent, unplug the printer and check all rectifier diodes and the filter capacitor as discussed in chapter 3. Replace any rectifier or filter components that you find defective or replace the supply; then re-test the printer.

Use your oscilloscope to observe any chopped dc output(s) from the switching IC. You should find high-frequency square waves (20 to 40 kHz) with an amplitude approximately equal to the preswitched dc level. This level would be about 38 V for Fig. 6-17. Set your oscilloscope to a time base of 5 to 10 μs/DIV, with a vertical sensitivity of at least 10 VOLTS/DIV. Adjust these settings as necessary to establish a clear trace.

If you do not read chopped dc, then either your switching regulator IC has failed, or one (or more) polarized filter capacitors have shorted. Unplug the printer and test each filter capacitor as shown in chapter 3. Replace any capacitor that appears to be shorted, then retest the supply. If all filter capacitors check correctly, replace the switching IC. Use care when desoldering the old IC to prevent any accidental PC board damage. Solder in an IC holder (if possible), then just plug in the new switching IC.

High-voltage supply troubleshooting

A high-voltage power supply is needed to provide the excitation that energizes the printer primary and transfer coronas, as well as the development roller. This book does not cover high-voltage supply background because of the safety dangers involved. The electrical stress on high-voltage components requires rather specialized components. For the purposes of this book, a defective high-voltage supply should be replaced rather than repaired.

Symptom 1 *Laser printing is too light or too dark.* As you saw in chapter 5, high voltage is a critical element of the laser printer image-formation system. High voltage is needed on the primary corona to provide a uniform charge to the drum surface. If high voltage drops off or fails completely, the resulting image will be light or nonexistent. The same thing is true of the development roller and transfer corona assemblies. If high voltage fails, toner will not transfer to the drum, and little (if any) toner that jumps to the drum will actually transfer to the page.

As a sanity check, try adjusting the contrast setting. An extremely low setting might appear to have these symptoms. If the problems persist with a high contrast setting, replace the high-voltage power supply.

Symptom 2 *Cannot control laser printer contrast.* In most cases, you should be able to optimize the printer contrast by adjusting the contrast control. When a new EP cartridge is installed, contrast should be slightly reduced to compensate for the abundant supply of fresh toner. As the EP cartridge is consumed, contrast can be increased to make up for the gradual loss of toner. If contrast cannot be adjusted, the contrast control knob might be defective. Because the contrast control is integrated with the high-voltage supply, replacing the high-voltage supply will replace the contrast control.

7
CHAPTER

Image-formation system

As discussed in chapter 5, electrophotographic (EP) printers (Fig. 7-1) use a complex combination of light, static electricity, heat, chemistry, and pressure, all guided by a complex *ECP* (electronic control package). There is no single part responsible for applying print—EP printers use a series of individual assemblies that make up its *IFS* (image-formation system). Because EP image formation uses a process rather than a print head, there are many more variable conditions that will affect the ultimate print quality and appearance. You have read about how EP printers work; this chapter presents detailed explanations and troubleshooting procedures specifically for EP (LED or laser) printers.

7-1
A Hewlett-Packard
LaserJet IV printer.

Hewlett-Packard Co.

Your printer *IFS* (image-formation system) is composed of eight major components as shown in Fig. 7-2: a photosensitive drum and developer assembly, cleaning blade, erase lamps, a primary corona assembly, a writing mechanism (a laser beam or LED array), transfer corona, static eliminator teeth, and a fusing assembly. Each element has to work properly to produce high-quality print. When a fault occurs in any of these areas, the resulting print will be adversely affected.

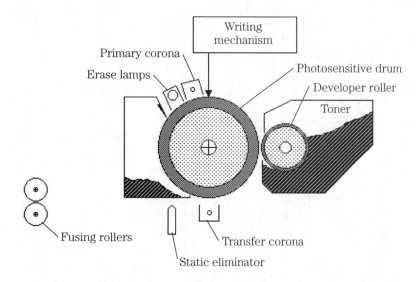

7-2 Components of an image-formation system.

System start-up problems

EP printers perform a self-test on start up to ensure that the ECP is active and responding normally. The self-test also checks communication between the printer and host computer. After the self-test is passed, the fusing assembly must reach upwards of 180°C within 90 seconds. When a self-test and warm up occur normally, the printer is generally ready to operate as long as paper and toner are detected. Unfortunately, printer start ups are not always so smooth. This part of the chapter details the symptoms and solutions for the most perplexing start-up problems.

Symptom 1 *Nothing happens when power is turned on.* You should hear the printer respond as soon as power is turned on. You should see a power indicator on the control panel (alphanumeric displays will typically indicate a self-test). You also should hear and feel the printer cooling fan(s) in operation. If the printer remains dead, there is probably trouble with the ac power. Check the ac line cord for proper connection with the printer and wall outlet. Also check the printer main ac fuse. When the ac and fuse check properly, there is probably a problem with the printer power supply. Refer to chapter 6 for power-supply troubleshooting.

If the printer fan(s) and power indicator operate, you can be sure the printer is receiving power. If the control panel remains blank, there might be a problem with the dc power supply or ECP. Follow the troubleshooting procedures of chapter 6 to check the printer dc power supply. When the power supply checks properly, the trouble is likely somewhere in the ECP or control panel assembly itself. Remove power from the printer and check the control panel cable. If there are no indicators at all on the control panel, replace the control-panel cable. If problems remain, try replacing the ECP. If you wish to troubleshoot the ECP in more detail, refer to the instructions of chapter 9. When only one or a few indicators appear on the control panel, try replacing the control panel cable. If problems remain, replace the control panel. If you wish to troubleshoot the control panel assembly in more detail, refer to chapter 9.

Symptom 2 *Your printer never leaves its warm-up mode. There is a continuous WARMING UP status code or message.* The initial self-test usually takes no more than 10 seconds from the time power is first applied. The fusing-roller assembly then must warm up to a working temperature, and is typically acceptable within 90 seconds from a cold start. At that point, the printer will establish communication with the host computer and stand by to accept data, so its WARMING UP code should change to an ON-LINE or READY code.

If the printer fails to go on-line, the problem is often a faulty communication interface, or a control-panel problem. Turn the printer off, disconnect its communication cable, and restore power. If the printer finally becomes ready without its communication cable, check the cable itself and its connection between the computer and printer. The cable might be faulty, or you might have plugged a parallel printer into the computer serial port (or vice versa). There also might be a faulty interface in your host computer or printer. Try a working printer (one that you know is working well) with the computer to ensure that the computer port is working correctly.

If the printer still fails to become ready with the communication cable disconnected, unplug the printer and check that the control-panel cables or interconnecting wiring are attached properly. Try reseating or replacing the control panel cable. Check the control panel to see that it is operating correctly. Try replacing the control panel. If problems persist, replace the ECP, which usually contains the control panel interface circuitry. Depending on the complexity of your particular printer, the interface/formatter might be a separate printed circuit plugged into the main logic board, or its functions might be incorporated right into the main logic board itself. If you wish to troubleshoot the control panel or ECP further, refer your troubleshooting to chapter 9.

Symptom 3 *You see a CHECKSUM ERROR message indicating a fault has been detected in the ECP program ROM.* During a self-test, the ECP will test its program ROM to see that it is working properly. This test is typically a checksum test of ROM contents. If the calculated checksum does not match with the checksum recorded on ROM, an error is generated. A checksum error usually indicates a failure of the ECP ROM device. Try the printer power off. Wait several minutes; then restore power. If the problem persists, replace the ECP or its interface/formatter module. If you want to attempt a more detailed repair, refer to chapter 9 and try replacing the ROM itself.

Symptom 4 *You see an error indicating communication problems between the printer and computer.* The printer and computer are not communicating properly. This symptom is typical in serial communication setups when baud rates or serial transfer protocols do not match exactly (refer to chapter 9 for detailed information on serial communication and transfer protocols). Check your serial communication cable first. Make sure that the cable is installed properly. Also make sure that it is the correct type and is wired properly for your printer. Keep in mind that pins 2 and 3 on the printer cable might need to be reversed for proper operation. If the pins must be reversed, use a *null modem* (available from almost any consumer electronics store) on the printer end of the cable. Also be aware of the cable length. Serial communication cables are typically limited to 15 meters (50 feet) , and Centronics (parallel) cables are limited to 3 meters (10 feet). Try a shorter cable if necessary.

There are five communication parameters that must match between the host computer and printer: *start* bits, *stop* bits, *data* bits, *parity* type, and *baud* rate. If any one of these parameters do not match, communication will not take place. At the printer end, there are usually DIP switch settings or control-panel key sequences that define each parameter (you might need to refer to the user's manual for your printer to determine how each parameter is set). At the computer end, you can usually set communication parameters directly through the application software that is doing the printing. Change parameters if necessary to set both printer and computer to the same parameters. Reboot the printer and computer.

Check to make sure that both the printer and computer are using the same serial flow control. Flow control is important because the host computer often must wait for the printer to catch up. XON/XOFF (software) and DTR/DSR (hardware) flow control are typically used. Adjust the printer or computer so that both use the same transfer protocol. Reboot the printer and computer.

If the printer still fails to operate (and you are certain that the computer is communicating properly), the communications port is probably defective. Try replacing the ECP or interface module.

Laser-delivery problems

Once the photosensitive drum gets a uniform electrical charge from the primary corona, a latent image is written to the drum surface. Writing is accomplished by discharging desired points along the drum surface with directed light. The classical method of writing is to scan a laser beam across the drum surface as shown in Fig. 7-3. This process is where the term *laser printer* comes from. The laser beam originates at a single stationary point in the printer, and is directed at a hexagonal mirror that is rotated at high speed. As a mirror rotates, the laser beam is directed (or *scanned*) through a compensating lens and across the drum surface. By turning the laser beam on and off corresponding to the presence or absence of dots along any one scan line, the desired bit image is written one line at a time. Typical laser printers can turn the laser on or off 300 times in any inch of scan line. When a scan line is completed, the drum rotates $\frac{1}{300}$ inch, and the next scan line begins. This process determines the printer overall resolution (300 × 300 dpi in this example).

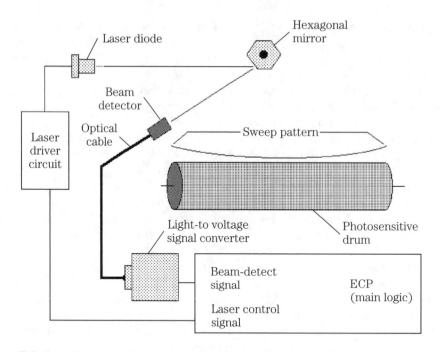

7-3 Scanning a laser beam across the drum surface.

Remember that the following symptoms and solutions are intended for laser printers because the laser and optical assemblies needed to implement a writing mechanism are particularly delicate. Remember that the error messages you see on your particular printer might not appear exactly as shown below—refer to your user's manual to compare your error code to the problem types shown.

Symptom 1 *You see a BEAM DETECTION error.* At the beginning of every scan line, the laser beam strikes an optical beam sensor. This registers the start of a new scan line and ensures that the data composing the new scan line is synchronized with the beam. Usually, the beam sensor is the only feedback that synchronizes the scan line and tells the laser printer that its laser is working. From time to time, unexpected variations in mirror rotation speed, age of the laser source, or the eventual buildup of dust or debris on laser optics might cause the beam to miss its sensor. These problems will cause the printer to register an error—that scan line will probably be missing on the printed page, but the error is usually recoverable.

If you find that the printer is registering random and occasional beam detection errors, check the printer optics. Printers with long service lifetimes might have accumulated enough dust or debris on the optics or beam sensor face to reduce beam power just enough to produce intermittent problems. Use a can of photography-grade compressed air and gently try blowing the dust away. If you cannot clear the contamination, use lint-free, photography-grade wipes lightly dampened with high-quality, photography-grade lens cleaners to wipe the lenses. Remember to be very gentle—take your time and let the wipe do the work. If you knock optics out of alignment, it will be virtually impossible to realign them again without factory service. If you mark the lens, it will be permanent, and you will need to have the lens replaced and realigned.

If the optics look good, the laser source might be failing. Older laser printers use a small, gas-filled helium-neon laser to produce the beam. With time, the helium-neon gas will escape and laser power will fall off. If the gas has escaped, the laser will have to be replaced and realigned (which requires factory service). Most current laser printers use solid-state laser diodes as the laser source. The semiconductor laser is combined with some switching and control circuitry and incorporated into a laser assembly with the scanning mirror and scanner motor as shown in Fig. 7-4. This assembly is called the *L/S* laser scanning assembly. Because the L/S assembly is completely prefabricated and designed as a replaceable component, it can be replaced outright with little risk of alignment problems.

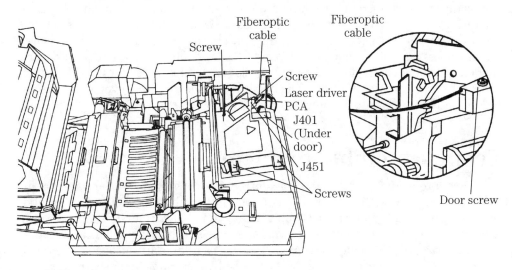

7-4 An installed laser/scanning assembly. Hewlett-Packard Co.

Use EXTREME caution whenever working with laser beams. Although the beam is invisible to the human eye and contains only a few milliwatts (mW) of power, looking directly at the beam (or reflections of the beam from other objects) can cause eye injury. Refer to warnings listed on the laser/scanning assembly for specific instructions.

Symptom 2 *You see a BEAM LOST error.* Although intermittent beam loss might result in a recoverable beam detection error, a prolonged loss of the laser beam (more than two seconds) will result in a more severe printer fault. There are many possible causes of laser loss. Begin by checking each voltage at the dc power supply. Most supplies mark their output levels, so you can check each output with a multimeter. If any dc level is low or absent, you can troubleshoot the supply as shown in chapter 6, or replace the dc supply outright. Check all connectors between the laser/scanning assembly and the printer to be sure that each is seated properly. Pay particular attention to the fiberoptic cable running from the L/S assembly to the ECP. This cable is the one that carries laser light to the detector. If this cable is loose, damaged, or disconnected, little or no laser signal will be delivered to the ECP.

There is a mechanical interlock (a shutter) that blocks the laser aperture whenever the printer case is opened. If the mechanical interlock becomes stuck or damaged, no laser beam will be available. Check the mechanical interlock carefully. You might have to remove the laser/scanning assembly to check the interlock. If the interlock is damaged, it will have to be replaced.

Usually, the solid-state laser diode in the laser/scanning assembly has failed, or the scanning mirror motor had stopped working. Replace the L/S assembly outright, and make sure to reattach each cable properly and completely.

Symptom 3 *You see a SCAN BUFFER error.* Remember that the laser beam must be turned on and off as the beam is scanned across the drum surface. Each dot across the image corresponds to the presence or absence of a bit in memory (called the *laser buffer*). As the laser sweeps across the drum, contents of the laser buffer are used to turn the laser on and off. If there is a problem with laser buffer memory, an error message will be generated. Usually, this is an intermittent fault that occurs randomly. Simply power down the printer, allow several minutes for it to clear, and reboot the printer again. If the problem persists, replace the ECP. If you wish to troubleshoot the ECP, refer to chapter 9.

Fusing-assembly problems

Once a toner image has been transferred from the drum to the page, toner must be permanently fixed (or fused) to the paper fiber. Fusing uses heat and pressure produced by a fusing assembly as shown in Fig. 7-5. In its simplest form, a fusing assembly is composed of five major parts: a heating roller, a pressure roller, a quartz heating lamp, a cleaning pad, and a temperature sensor. Although there is certainly other hardware in the assembly, these are the parts that are actually doing the work.

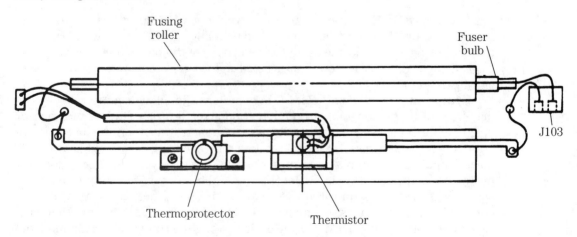

7-5 A fusing/sensor assembly. Hewlett-Packard Co.

Paper passes between the two rollers. The bottom roller simply provides pressure against the top roller. The top roller is heated from the inside by a long, thin quartz lamp powered by your printer ac supply. Although the top roller is made with a material that prevents toner from sticking, some toner particles will inevitably stick. If those particles are not cleaned away, they could stick on later parts of the page and cause problems. A cleaning pad is included in the fusing assembly to wipe off any toner particles from the heating roller. Many manufacturers provide an easily replaceable cleaning pad that can be changed when a new toner cartridge is installed. Finally, a temperature sensor (usually a thermistor) is included with the assembly to regulate the heat being applied to the page.

Symptom *You see a SERVICE error indicating a fusing malfunction.* Fusing is important to the successful operation of any EP printer. Toner that is not fused successfully remains a powder or crust that can flake or rub off on your hands or other pages. Main logic interprets the temperature signal developed by the thermistor and modulates ac power to the quartz lamp. Three conditions will generate a fusing malfunction error: (1) fusing-roller temperature falls below about 140°C, (2) fusing-roller temperature climbs above 230°C, or (3) fusing-roller temperature does not reach 165°C in 90 seconds after the printer is powered up. Your particular printer might use slightly different temperature and timing parameters. When such an error occurs, your first action should be to power down the printer and reboot. Note that a fusing error will often remain with a printer for 10 minutes or so after it is powered down, so be sure to allow plenty of time for the system to cool before rebooting after a fusing error.

If the error persists, power down the printer and examine the installation of your fusing assembly. Check to see that all wiring and connectors are tight and seated properly. An ac power supply is often equipped with a fuse or circuit breaker that protects the printer. If this fuse or circuit breaker is open, replace your fuse or reset your circuit breaker, then re-test the printer. Remember to clear the error, or allow enough time for the error to clear by itself. If the fuse or breaker trips again during re-test, you have a serious short circuit in your fusing assembly or ac power supply. You can attempt to isolate the short circuit, or simply replace your suspected assemblies—ac power supply first, then the fusing assembly.

Unplug the printer and check your temperature sensor thermistor by measuring its resistance with a multimeter. At room temperature, the thermistor should read about 1 kΩ (depending on the particular thermistor). If the printer has been at running temperature, thermistor resistance might be much lower. If the thermistor appears open or shorted, replace it with an EXACT replacement part and re-test the printer.

A thermal switch (sometimes called a *thermoprotector*) is added in series with the fusing lamp. If a thermistor or main-logic failure should allow temperature to climb out of control, the thermal switch will open and break the circuit once it senses temperatures over its preset threshold. This switch protects the printer from severe damage—and possibly a fire hazard. Unplug the printer, disconnect the thermal switch from the fusing lamp circuit, and measure its continuity with a multimeter. The switch should normally be closed. If you find an open switch, replace it. Check the quartz lamp next by measuring continuity across the bulb itself. If you read an open circuit, replace the quartz lamp (or the entire fusing assembly). Be sure to secure any disconnected wires.

If the printer still does not reach its desired temperature, or continuously opens the thermal switch, there is probably a fault in the ECP. Try replacing the ECP. If you wish to troubleshoot the ECP, refer to chapter 8 for a discussion of sensors, or chapter 9 for detailed information about the ECP.

Image-formation problems

There are many variables at work in the formation of an electrophotographic image, so even though the printer might be operating within safe limits where no error messages are generated, the printed image might not be formed properly. Although print quality is always a subjective decision, there are certain physical characteristics in EP printing that signal trouble in image formation. It is virtually impossible to define every possible image formation problem, but this part of the chapter illustrates a broad range of basic symptoms that can tell you where to look for trouble.

Symptom 1 *Pages are completely blacked out, and might appear blotched with an undefined border (Fig. 7-6).* Unplug the printer, remove the EP cartridge, and examine its primary corona wire. Remember from chapter 5 that a primary corona applies an even charge across a drum surface. This charge readily repels toner—except at those points exposed to light by the writing mechanism that attract toner. A failure in the primary corona will prevent charge development on the drum. As a result, the entire drum surface will attract toner (even if your writing mechanism works perfectly). If the entire surface attracts toner, the image will be totally black. If you find a broken or fouled corona wire, clean the wire or replace the EP cartridge.

7-6
Printed page is blacked out.

If your blacked-out page shows print with sharp, clearly defined borders, your writing mechanism might be running out of control. LEDs in a solid-state print bar or laser beam might be shorted in an ON condition, or receiving erroneous data bits from its control circuitry (all logic 1s). In this example, the primary corona is working just fine, but a writing mechanism that is always on will effectively expose the entire drum and discharge whatever charge was applied by the primary corona. The net result of attracting toner would be the same, but whatever image is formed would probably appear crisper—more deliberate.

Your best course here is simply to replace the ECP. If you choose to troubleshoot the ECP, use your oscilloscope to measure the data signals reaching your writing mechanism during a print cycle. You should find a semirandom square wave representing the 1s and 0s composing the image. If you find only one logic state, troubleshoot your main logic and driving circuits handling the data. If data entering the writing mechanism appears normal, replace your writing mechanism.

Symptom 2 *Print is very faint (Fig. 7-7).* Before attempting anything else, try adjusting the printer contrast control. If that fails to help, unplug the printer, remove the EP cartridge, and try redistributing toner in the cartridge. Your user's manual probably offers preferred instructions for redistributing toner. Keep in mind that toner is largely organic—as such, it has only a limited shelf and useful life. If redistribution temporarily or partially improves the image, or if the EP cartridge has been in service for more than six months, replace the EP cartridge. If you are using a paper with a moisture content, finish, or conductivity that is not acceptable, image formation might not take place properly.

7-7
Printed page is very faint.

Check your transfer corona. The transfer corona applies a charge to paper that pulls toner off the drum. A weak transfer corona might not apply enough charge to attract all the toner in a drum image. This low charge can result in very faint images. Unplug the printer, allow ample time for the high-voltage power supply to discharge completely, then inspect all wiring and connections at the transfer corona. If the monofilament line encircling the transfer corona is damaged, replace the transfer corona assembly, or attempt to rethread the monofilament line. If faint images persist, repair or replace the high-voltage power supply assembly.

Finally, check the drum ground contacts to be sure that they are secure. Dirty or damaged ground contacts will not readily allow exposed drum areas to discharge. As a result, very little toner will be attracted and only faint images will result.

Symptom 3 *Print appears speckled (Fig. 7-8).* Your first step should be to turn off printer power and check the cleaning pad on the fusing roller. A pad that is old or worn will not wipe the roller properly and should be replaced. Replace the cleaning pad if necessary and re-test the printer.

7-8
The print has a speckled or dirty appearance.

If the cleaning pad checks out, speckled print is probably the result of a fault in your primary corona grid. A grid is essentially a fine wire mesh between the primary corona and drum surface. A constant voltage applied across the grid serves to regulate the charge applied to the drum to establish a more consistent charge distribution. Grid failure will allow much higher charge levels to be applied unevenly. The higher charge results in dark splotches in the print. Because the primary grid assembly is part of the EP cartridge, replace the EP cartridge and re-test the printer. If speckled print persists, repair or replace the high-voltage power supply assembly.

Symptom 4 *There are one or more vertical white streaks in the print (Fig. 7-9).* Begin by checking your toner level. Toner might be distributed unevenly along the cartridge length. Unplug the printer, remove the EP cartridge, and redistribute the toner. Follow your manufacturer's recommendations when handling the EP cartridge. If this improves your print quality (at least temporarily), you know that the EP cartridge must be replaced soon.

Next, examine your transfer corona for areas of blockage or extreme contamination. Such faults would prevent the transfer corona from generating an even

```
One o    re vertic:  white
streak   pear in tl  ə
printe   ge.  This   ype
of syr   n is usua   y due
to a l   r uneven    :oner
level.

Try r    ing the E
cartri   irst  If tl  it fails
there    · be a blo  ‹ in
the o]   or laser/:  anning
assem      Try clea  ling
the a:   oly
```

7-9
The print contains one or more
vertical white streaks.

charge along its length—corrosion acts as an insulator that reduces the corona electric field. Uncharged page areas will not attract toner from the drum, so those page areas will remain white. Clean the transfer corona very carefully with a clean cotton swab. If your printer comes with a corona cleaning tool, use that instead. When cleaning, be sure to avoid the monofilament line wrapped around the transfer corona assembly. If the line breaks, you will need to rewrap it or replace the entire transfer corona assembly.

Check the optics for any accumulated dust or debris that could block out sections of light. Because ES drums are only scanned as fine horizontal lines, it would take little more than a fragment of debris to block light through a focusing lens. Gently blow off any dust or debris with a can of high-quality, photography-grade compressed air available from any photography store. For stains or stubborn debris, clean the afflicted lens gently with high-quality, lint-free wipes and lens cleaner from any photography store. Be very careful not to dislodge the lens from its mounting. Never blow on a lens or mirror yourself! Breath vapor and particles can condense and dry on a lens to cause even more problems in the future.

Symptom 5 *Right-hand text appears missing or distorted (Fig. 7-10).* Usually, this is simply a result of low toner in your EP cartridge. If any area of the development roller receives insufficient toner, it will result in very light or missing image areas. Unplug the printer, remove the EP cartridge, and redistribute the toner. Follow your manufacturer's recommendations for toner redistribution. If you see an improvement in image quality (at least temporarily), replace the EP cartridge.

Examine the mountings that support your writing mechanism. The mechanism, along with its associated optics, is usually mounted to a removable cover. Make sure that the writing mechanism (laser, LED, or LCS) is mounted correctly, and that its cover is closed completely. If the writing mechanism is not mounted correctly, scan lines might not be delivered to the proper drum locations. Try replacing the writing mechanism.

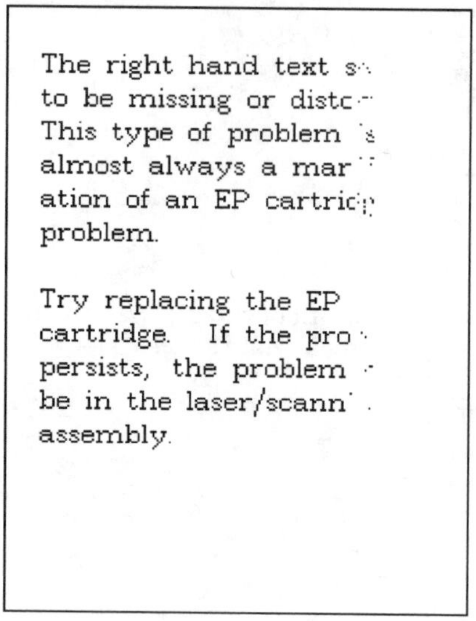

The right hand text s
to be missing or distc
This type of problem s
almost always a mar
ation of an EP cartric
problem.

Try replacing the EP
cartridge. If the pro
persists, the problem
be in the laser/scann
assembly.

7-10
Right-hand print appears missing
or distorted.

If you are using a laser writing mechanism, pay special attention to the installation and alignment of the laser/scanning assembly. If the assembly is not installed with the correct alignment, the scanning beam might start or stop at different points along the drum. An end portion of the image might be distorted or missing. Reseat or replace any incorrectly positioned laser/scanning assembly. If you are working with an older laser printer, check the alignment of the scanning mirror.

Symptom 6 *You consistently encounter faulty image registration* (Fig. 7-11). Paper sheets are drawn into the printer by a pickup roller, and held by a set of registration rollers until a drum image is ready to be transferred to paper. Under normal circumstances, the leading edge of paper will be matched (or *registered*) precisely with the beginning of a drum image. Poor paper quality, mechanical wear, and paper path obstructions can all contribute to registration problems.

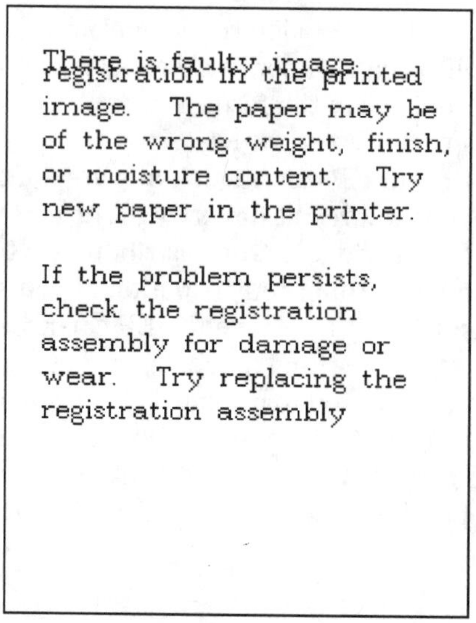

There is faulty image
registration in the printed
image. The paper may be
of the wrong weight, finish,
or moisture content. Try
new paper in the printer.

If the problem persists,
check the registration
assembly for damage or
wear. Try replacing the
registration assembly

7-11
The image does not register
properly.

Begin by inspecting your paper and paper tray assembly. Unusual or specialized paper might not work properly in your paper transport system (this also can lead to PAPER JAM errors as discussed in chapter 8). Check the paper specifications for your printer listed in your user's manual. If you find that your paper is nonstandard, try about 50 sheets of standard-bond xerographic paper and re-test the printer. Because paper is fed from a central paper tray, any obstructions or damage to the tray can adversely affect page registration (or even cause paper jams). Examine your tray carefully. Correct any damage or restrictions that you might find, or replace the entire tray outright.

If registration is still incorrect, it usually suggests mechanical wear in your paper feed assembly. Check the pickup roller assembly first. Look for signs of excessive roller wear. Remove your printer housings to expose the paper transport system. You will have to defeat any housing interlock switches, and perhaps the EP cartridge sensitivity switches. Perform a self-test and watch paper as it moves through the printer. The paper pickup roller should grab a page and move it about 3 inches or more into the printer before registration rollers activate. If the pickup roller clutch solenoid turns on, but the pickup roller fails to turn immediately, your pickup assembly is worn out. The recommended procedure is simply to replace the pickup assembly, but you might be able to adjust the pickup roller or clutch tension to improve somewhat printer performance.

Another common problem is wear in the registration roller assembly. If this set of rollers does not grab the waiting page and pull it through evenly at the proper time, the image on the paper might not be correct. As you initiate a printer self-test, watch the action of your registration rollers. The rollers should engage immediately after the pickup roller stops turning. If the registration clutch solenoid activates, but paper does not move immediately, your registration roller assembly is worn out. The recommended procedure simply is to replace the registration assembly, but you might be able to adjust torsion spring tensions to somewhat improve printer performance.

Pay particular attention to the components in your drive train assembly. Dirty or damaged gears can jam or slip, which leads to erratic paper movement and faulty registration. Clean your drive train gears with a clean, soft cloth. Use a cotton swab to clean gear teeth and tight spaces. Remove any objects or debris that might block the drive train, and replace any gears that are damaged.

Symptom 7 *You encounter horizontal black lines spaced randomly through the print (Fig. 7-12).* Remember that black areas are ultimately the result of light striking the drum. If your printer uses a laser writing mechanism, a defective or improperly seated beam detector could send false scan timing signals to the main logic. The laser would make its scan line while main logic waits to send its data. At the beginning of each scan cycle, the laser beam strikes a detector. The detector carries laser light through an optical fiber to a circuit that converts light into an electronic logic signal that is compatible with main logic. Main logic interprets this *beam-detect signal* and knows the scanner mirror is properly aligned to begin a new scan. Main logic then modulates the laser beam on and off corresponding to the presence or absence of dots in the scan line.

Positioning and alignment are critical here. If the beam detector is misaligned or loose, the printer motor vibrations might cause the detector to occasionally miss the

There are horizontal black
lines spaced randomly
through the printed page.
This kind of intermittent
problem is usually due to
a problem with the laser
beam sensor or optical
cable. Try replacing the
laser/scanning assembly

7-12
There are horizontal black lines in
the print.

beam. Main logic responds to this by activating the laser full-time in an effort to synchronize itself again. Reseat or replace the beam detector and optical fiber.

A loose or misaligned laser/scanning assembly also can cause this type of problem. Vibrations in the mirror might occasionally deflect the beam around the detector. Realign, reseat, or replace the laser/scanning assembly.

Symptom 8 *Print is slightly faint.* Print that is only slightly faint does not necessarily suggest a serious problem. There are a series of simple checks that can narrow down the problem. Check the print contrast control dial. Turn the dial to a lower setting to increase contrast (or whatever darker setting there is for your particular printer). If the contrast control has little or no effect, your high-voltage power supply is probably failing. Replace your high-voltage power supply.

Check your paper supply. Unusual or specially coated paper might cause fused toner images to appear faint. If you are unsure about the paper currently in the printer, insert a good-quality, standard-weight xerographic paper and test the printer again.

Check your toner level. Unplug the printer, remove the EP cartridge, and redistribute toner. Follow all manufacturer's recommendations when it comes to redistributing toner. The toner supply might just be slightly low at the developing roller.

Unplug your printer and examine the EP cartridge sensitivity switch settings. These microswitches are actuated by molded tabs attached to your EP cartridge. This tab configuration represents the relative sensitivity of the drum. Main logic uses this code to set the power level of its writing mechanism to ensure optimum print quality. These switches also tell main logic whether an EP cartridge is installed at all. If one of these tabs are broken, or if a switch has failed, the drum might not be receiving enough light energy to achieve proper contrast. Check your sensitivity switches as outlined for a NO EP CARTRIDGE error shown in chapter 8.

Over time, natural dust particles in the air will be attracted to the transfer corona and accumulate there. The dust eventually causes a layer of debris to form on the wire. This type of accumulation cuts down on transfer corona effectiveness, which places less of a charge on paper. Less toner is pulled from the drum, so the resulting image appears fainter. Unplug the printer, allow time for the high-voltage power supply to discharge, then gently clean the transfer corona with a clean cotton swab or corona cleaning tool. Be very careful not to break the monofilament line wrapped about the transfer corona assembly. If this line does break, the transfer corona assembly will have to be rewrapped or replaced.

Symptom 9 *Print has a rough or suede appearance (Fig. 7-13).* Usually, suede print is the result of a serious failure in the main logic system (ECP). The writing mechanism is being allowed to turn on and off randomly during its scanning cycles. This type of symptom is dominant in laser printers where a faulty laser driver can allow the beam to act erratically. Your best course is usually to replace the ECP outright. If you wish to troubleshoot the ECP, use your oscilloscope to trace the print data signal from the writing mechanism into main logic circuitry. You will need a schematic diagram of your printer for this troubleshooting.

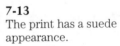

7-13
The print has a suede appearance.

Symptom 10 *Print appears smeared or improperly fused (Fig. 7-14).* Temperature and pressure are two key variables of the EP printing process. Toner must be melted and bonded to a page to fix an image permanently. If fusing temperature or roller pressure is too low during the fusing operation, toner might remain in its powder form. Resulting images can be smeared or smudged with a touch.

Perform a simple fusing check by running several continuous self-tests (the printer does not have to be disassembled for this). After about ten printouts, place the first and last printout on a firm surface and rub both surfaces with your fingertips. No smearing should occur. If your fusing level varies between pages (one page might smear, and another might not), clean the thermistor temperature sensor and repeat this test. Remember to wait 10 minutes or so and unplug the printer before working on the fusing assembly. If fusing performance does not improve, replace the

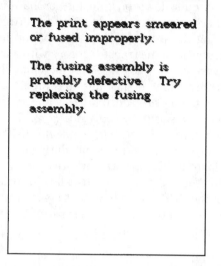

The print appears smeared or fused improperly.

The fusing assembly is probably defective. Try replacing the fusing assembly.

7-14
Print smears or appears improperly fused.

thermistor and troubleshoot its signal conditioning circuit. If smearing persists, replace the fusing assembly and cleaning pads.

Static teeth just beyond your transfer corona are used to discharge the paper once toner has been attracted away from the drum. This helps paper to clear the drum without being attracted to it. An even charge is needed to discharge paper evenly, otherwise, some portions of the page might keep a local charge. As paper moves toward the fusing assembly, remaining charge forces might shift some toner resulting in an image that does not smear to the touch, but has a smeared or pulled appearance. Examine the static discharge comb once the printer is unplugged and discharged. If any of its teeth are bent or missing, replace the comb.

A cleaning pad rubs against the fusing roller to wipe away any accumulations of toner particles or dust. If this pad is worn out or missing, contamination of the fusing roller can be transferred to the page, resulting in smeared print. Check your cleaning pad in the fusing assembly. Worn out or missing pads should be replaced immediately.

Inspect your drive train for any gears that show signs of damage or excessive wear. Slipping gears could allow the EP drum and paper to move at different speeds. The different speeds can easily cause portions of an image to appear smudged—such areas would appear bolder or darker than other portions of the image. Replace any gears that you find to be defective. If you do not find any defective drive train components, try replacing the EP cartridge.

Finally, a foreign object in the paper path can rub against a toner powder image and smudge it before fusing. Check the paper path and remove any debris or paper fragments that might be interfering with the image.

Symptom 11 *Printed images appear to be distorted (Fig. 7-15). Distortion* is at best a vague term when applied to printed images, but for the purpose of this symptom, you might see one of two types of distortion: image size distortion, and scanning distortion.

Image size distortion is indicated when characters appear too large or too small in the vertical direction. Large (or *stretched*) characters suggest that some portion of the pickup or registration assemblies is failing, or that there is some obstruction in

**Image distortion is shown
by characters which
appear too large or small
in the vertical direction**

Scanning distortion is
typically the result of
a laser/scanning
assembly problem

7-15
The image appears distorted.

the paper path causing excessive drag on the paper. Characters that are too small (or *squashed*) suggest a main motor problem—it might be moving the drum too fast.

Examine your pickup and registration assemblies for signs of unusual wear and replace any rollers or other mechanics that appear worn or damaged. Also inspect your EP cartridge. If the cartridge is very new or very old, it might be worth trying a replacement cartridge. If characters appear compressed, check your main motor and motor drive signals. Finally, look for any debris or obstructions that might interfere with drive train operation. Remove any obstructions immediately.

Scanning distortion (found in laser printers) is indicated by wavy, irregularly-shaped characters. This wavy distortion also can be seen in the page margins. Usually, a marginal scanning motor causes minor variations in scanning speed (the motor speeds up and slows down erratically). For example, consecutive horizontal scan lines will not appear parallel. If all connectors and interconnecting wiring to the laser/scanning assembly appear correct, simply replace the laser/scanning assembly outright. If you are working on an older laser printer, you can probably replace the mirror scanner assembly as a separate unit.

Symptom 12 *Print shows regular or repetitive defects (Fig. 7-16).* Repetitive defects are problems that occur at regular intervals along a page (as opposed to random defects) and are most often the result of roller problems. Rollers have fixed circumferences, so as paper moves through the printer, any one point on a roller might reach a page several times. For example, if a drum has a circumference (not a diameter) of 3 inches, any one point on the drum will reach a standard 8.5 inches × 11 inches page up to 3 times. If the drum is damaged or marked at that point, those imperfections will repeat regularly in the finished image.

Many repetitive defects take place in the EP cartridge that contains the photosensitive drum and developing roller. A typical drum has a circumference of about 3.75 inches. Defects that occur at that interval can often be attributed to a drum defect. A developing roller has a circumference of about 2 inches, so problems that repeat every 2 inches are usually associated with the developing roller. In either case, replace the EP cartridge and re-test the printer.

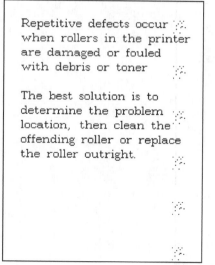

Repetitive defects occur
when rollers in the printer
are damaged or fouled
with debris or toner

The best solution is to
determine the problem
location, then clean the
offending roller or replace
the roller outright.

7-16
Repetitive defects appearing in
the print.

A fusing roller has a circumference of about 3 inches. Image marks or defects at that interval suggest a dirty or damaged fusion roller. Unplug the printer, allow at least 10 minutes for the fusion assembly to cool, then gently clean the fusion rollers. If you find that the fusion rollers are physically damaged or can't clean them effectively, replace the fusing-roller assembly.

Any roller that is fouled with debris or toner particles can contribute to a repeating pattern of defects. Make sure to examine each of your rollers carefully. Clean or replace any roller that you find to be causing marks.

Symptom 13 *The page appears completely black except for horizontal white stripes (Fig. 7-17).* This symptom indicates an intermittent loss of the laser beam either in the laser/scanning assembly itself or in the fiberoptic detector cable.

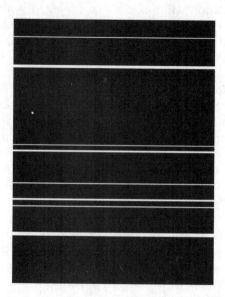
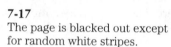

7-17
The page is blacked out except
for random white stripes.

If the printer cannot detect the laser beam, the laser will fire full-duty as the printer tries to re-establish synchronization. This will expose the drum and result in black print. The white lines indicate that synchronization is briefly restored. Your first action should be to check the fiberoptic cable between the laser/scanning assembly and the ECP. If the cable appears to be connected properly, it should be replaced. Note that you also might see a BEAM DETECTION error with this type of problem. If the problem persists, the trouble is probably due to a defect in the laser/scanner assembly. Try replacing the laser/scanner assembly outright.

Symptom 14 *The image appears skewed (Fig. 7-18).* Image skew is usually the result of a problem in the paper path—something is happening as paper travels through the printer. Begin by checking the paper tray. Make sure that the right type and weight of paper is installed properly in the tray (the tray might have too much paper). Remove and reseat the paper tray to be sure that it is inserted evenly and completely. Try switching trays with another compatible printer.

```
The image seems to be
skewed at one or more
points along the page.

This is usually the result
of some sort of physical
obstruction in the paper
handling path.

Check the paper path
and clear any obstructions
that you may find.
```

7-18
The image appears skewed at one or more points.

If the paper and tray both appear intact, unplug the printer, and open the outer covers. Inspect the paper pickup roller for signs of unusual or uneven wear. If the pickup roller is uneven, paper will walk (or *skew*) before it reaches the registration rollers. When you find a worn or damaged paper pickup roller, you might be able to adjust the roller tension mechanism to compensate, but you are usually best advised simply to replace the paper pickup assembly.

If the paper pickup assembly looks good, check the registration roller assembly next. Registration rollers hold the page in place until the latent drum image is positioned properly for the page. As with the paper pickup, a worn or damaged registration assembly will allow the page to skip or walk as it is being passed through the printer. Note that the registration assembly uses torsion springs to ensure even pressure across the rollers. A missing or defective spring can result in uneven pressure. If you find uneven pressure, you can readjust the torsion springs, but your best course (especially for older printers) is to replace the registration roller assembly outright.

If the problem persists, there is probably an obstruction somewhere in the paper path. You will need to examine the paper path very carefully to look for obstructions. Stuck labels and paper fragments are typical causes. Gently clear any obstructions that you find—be very careful not to damage any rollers or mechanisms in the process of clearing an obstruction.

Symptom 15 *The image is sized improperly.* The EP printer control circuitry sizes an image based on the paper in the tray. Printers use a series of microswitches that are actuated by a specially shaped tab attached to each tray, and each tray uses a differently shaped tab to actuate these *tray-detect* switches in a unique sequence. Make sure that each tray is fitted with the proper tab for the tray being used.

If the tray is fitted with the proper tab, there might be a problem with the printer tray-detect microswitches. Either the switches or ECP might be at fault. If you wish to troubleshoot the paper size microswitches, refer to the procedures in chapter 8.

Symptom 16 *There are vertical black streaks in the image (Fig. 7-19).* This symptom is indicated by one or more dark vertical lines. Each line might be a different width, but the width of each line will remain constant throughout the length of the page. Usually, you will find that the primary corona is dirty. Remember from chapter 5 that any gunk or debris that accumulates on the primary corona will prevent an even charge distribution. Fortunately, this problem can be rectified simply by cleaning the primary corona wire. Use care when cleaning the primary corona. If it breaks, the EP cartridge will have to be replaced. If the primary corona is clean but the problem persists, there might be a manufacturing problem with the EP cartridge itself. Nicks or manufacturing defects around the development roller can allow much more toner in proximity of the drum. Try a new EP cartridge.

```
Vertical black streaks
appear in the image.  This
type of symptom almost
always indicates that the
primary corona is dirty.

Clean the primary corona
carefully.
```

7-19
Vertical black streaks appear in the image.

For LED printers, there might be a problem with the scan line buffer memory. A scan line buffer holds the individual pixels that will ultimately appear in the line. If a new EP cartridge does not solve the problem, try replacing the ECP to exchange the scan line memory. If you wish to try troubleshooting the ECP in detail, refer to the procedures of chapter 9.

8
CHAPTER

Mechanical systems

More is needed than just good electronics and a fancy case to make a first-class printer. A series of tightly integrated mechanical systems are needed to perform the variety of physical tasks that every laser printer (Fig. 8-1) must do. The most obvious mechanism is the paper transport system responsible for moving paper through the printer, but the image formation system also needs mechanical force to function. Laser printers need a well-regulated scanner motor assembly that can direct a laser beam across a light-sensitive drum surface. You have already read about a selection of important mechanical parts in chapter 2, but this chapter shows you how to deal with mechanical problems.

8-1 A Tandy LP800 laser printer.

Paper problems

Before you jump right into paper detection and handling symptoms, you should have a solid understanding of how paper is handled in a typical EP printer. Paper in a paper tray is loaded into the printer. A *paper-detect* sensor makes sure that paper is available in the tray, and the tray itself actuates a series of *tray size* micro-switches. Each tray size actuates a different sequence of switches that tell the ECP just which tray size is now installed, which allows the printer to automatically size the image according to the paper tray being used.

When a printing cycle begins, the main motor turns, which causes mechanical linkages to turn the EP drum, fusing rollers, and any feed rollers that carry the paper along. The only two mechanisms that do not turn are the *paper pickup roller* and the *registration rollers*. Their actions are regulated by solenoid-driven clutches that remain open. When the printing cycle is ready to receive a page, the pickup roller clutch engages. The notched pickup roller grabs the top page and draws it into the printer about 7.5 centimeters (about 3 inches), which is about the circumference of the pickup roller. After one turn, the pickup-roller clutch disengages, and the page rests against the registration rollers. A rubber *separation pad* just below the pickup roller prevents more than one page at a time from entering the printer.

When the developed drum image is properly positioned in relation to the page, the registration roller clutch engages and starts the page into the image formation system. Once the registration rollers start, they will remain engaged until the paper has exited the printer. Feed rollers guide the page while the latent image is transferred to it, then gently transfer the page to the fusing assembly where the toner image is fixed. As the paper emerges from the fusing rollers, a *paper exit* sensor is actuated. When the page has left the printer, the sensor assumes its original condition, and the printer ECP knows that the page has left the printer. The main motor can now stop (or a new page can enter the printer for another cycle).

Symptom 1 *You find a PAPER OUT message.* When the printer generates a PAPER OUT message, either the paper tray is empty or the paper tray has been removed. When a paper tray is inserted, a series of metal or plastic tabs contact a set of microswitches as shown in Fig. 8-2. The presence or absence of tabs will form a

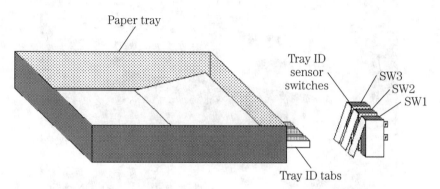

8-2 Paper tray ID (identification) switch configuration.

code that is unique to that particular paper size. Microswitches are activated by the presence or tabs. Main logic interprets this paper type code, and knows automatically what kind of media (paper, envelopes, etc.) that it is working with. Table 8-1 shows a typical paper code.

Table 8-1.
Typical tray switch configurations

Tray type	SW1	SW2	SW3
Executive	1	1	1
A4	1	1	0
Legal	0	0	1
Envelope	0	1	1
Letter	1	0	0
No tray	0	0	0

1 = on (engaged).

0 = off (disengaged).

The presence of paper is detected by a mechanical sensing lever as shown in Fig. 8-3. When paper is available, a lever rests on the paper. A metal or plastic shaft links this lever to a thin plastic flag. While paper is available, this flag is clear of the paper-out sensor. If the tray becomes empty, this lever falls through a slot in the tray, which

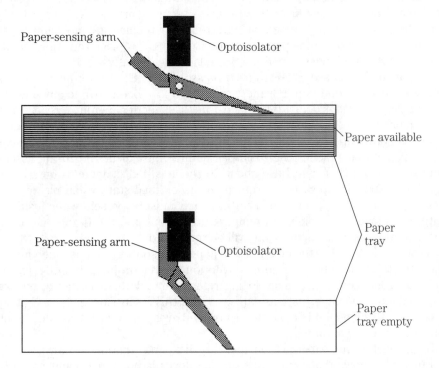

8-3 Operation of a paper-sensing arm.

rotates its flag into the paper-out sensor, which shows that paper is exhausted. The paper-out sensor is usually mounted on an auxiliary PC board (known as the *paper-control* board).

Begin your check by removing the paper tray. Be sure that there is paper in the tray, and that any ID (identification) tabs on the tray are intact—especially if you have just recently dropped the tray. Re-insert the filled paper tray carefully and completely. If the PAPER OUT message continues, then there is either a problem with your paper tray ID microswitches, paper sensing lever, or the paper-out optoisolator.

You can check the paper ID microswitches by removing the paper tray and actuating the paper sensing lever by hand (so the printer thinks that paper is available). Refer to Table 8-1 and actuate each switch in turn using an eraser of a pencil. Actuate one switch at a time and observe the printer display. The PAPER OUT error should go away whenever at least one microswitch is pressed. If the error remains when a switch is pressed, that switch is probably defective. Unplug the printer and use your multimeter to check continuity across the suspect switch as you actuate it. Replace any defective switch or switch assembly. Inspect the paper-out lever and optoisolator next.

When paper is available, the paper-out lever should move its plastic flag clear of the optoisolator. When paper is empty, the lever should place its flag into the optoisolator slot. Note: this logic might be reversed depending on the particular logic of the printer. Actuate the paper-out lever by hand and see that it moves freely and completely. This check confirms that the paper sensing arm works properly. If you see the lever mechanism jammed or bent, repair or replace the mechanism. If the problem persists, replace the paper-out optoisolator (or replace the paper handling PC board). If the problem still continues, replace the defective ECP.

Symptom 2 *You see a PAPER JAM message.* The EP printer must detect and report a paper-jam condition. For most printers, a *jam* occurs when any one of the four following situations do not take place. First, a sheet of paper must reach the fusing assembly within some amount of time after the printing cycle starts (usually under 30 seconds). Paper reaching the fusing assembly actuates the paper exit sensor. Second, paper that reaches the fusing assembly, must clear the fusing assembly within some amount of time (also about 30 seconds, but it depends on the paper size). The actuated paper sensor returns to its original state when paper passes. Third, paper that is present in the fusing assembly must be present when fusing temperature is above some minimum temperature (about 150°C). If any one of these three conditions is false, a paper jam will be registered.

As shown in Fig. 8-4, the electrostatic paper transport system is much more sophisticated than those used in more conventional serial or line printers. As a result of this additional complexity, main-logic circuitry must detect whether or not paper enters and exits the paper path as expected. For this discussion, assume that a jam can occur in three general areas: the paper-feed area, the registration and transfer area, and the fusing area.

Begin by checking paper in the paper tray. If a jam condition is shown, but there is no paper, it shows that your paper sensing lever is not functioning properly. It might be broken, bent, or jammed. When there is ample paper available, take a mo-

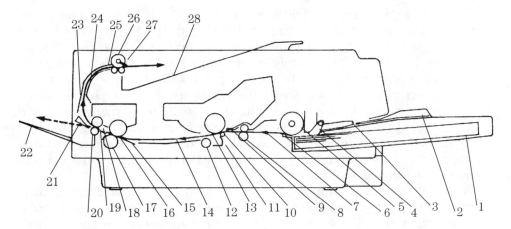

8-4 A cross-sectional diagram showing the paper path. <small>Hewlett-Packard Co.</small>

ment to be sure that paper is the right size, texture, and weight for your printer. Unusual or special paper might not be picked up reliably. If you are uncertain as to the correct paper type, remove it and insert a quantity of good-quality xerographic paper. This type of paper usually has the weight and texture characteristics ideal for ES printing. If the error continues, look at potential jam locations—paper-feed area, registration/transfer area, and exit area—as described below.

Paper-feed area The paper-feed area consists of the paper tray (and paper), pickup roller mechanical assembly, and electromechanical clutch as shown in Fig. 8-5. If paper is not reaching your registration rollers, the trouble is probably in this area.

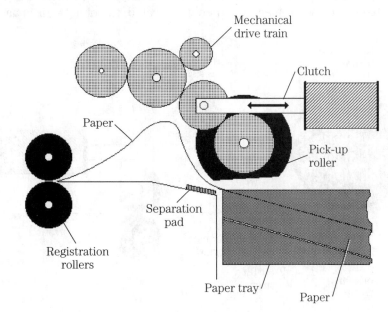

8-5 Simplified diagram of a paper pickup/feed mechanism.

Inspect your paper tray carefully. Although the tray might seem foolproof, it actually plays an important role in paper feed. If the plastic tray housing is cracked or damaged, replace it with a new tray and re-test the printer. Note the movable metal plate in the tray bottom. This lift mechanism keeps paper positioned against the pickup roller at all times. Remove the tray and paper. Make sure that this plate can move freely—replace your paper tray if it does not. Observe this lift plate as you insert it into the printer. The printer should lift this plate up as the tray is inserted. If this does not happen, repair or replace the printer lift mechanism assembly. Add some fresh paper (50 sheets or so) and gently insert the paper tray. Be sure to insert the tray fully and squarely. If there is any obstruction (or the tray does not seat squarely), find and remove the obstruction.

Next, make sure that your main motor is functioning. Keep in mind that the main motor drives all rollers in the printer, as well as the photosensitive drum. If this motor has failed, paper will not be drawn into the printer at all. You can observe the main motor and its gear-train assembly by turning the printer on, opening an access cover, defeating the associated interlock switch (if any), defeating EP cartridge sensitivity switches, and initiating a printer self-test. If the fusing roller temperature is above its lower temperature limit, you should see motor operation immediately. If the main motor fails to operate, replace the main motor driver circuit or ECP as required. If the main motor continues to malfunction, replace the main motor assembly.

The main motor turns when a print signal is first generated in main logic. It continues turning as long as there is paper in the feed path. However, even though the main motor supplies the force to operate every roller, pickup and registration rollers are operated only briefly during each print cycle. An electromagnetic clutch (also shown in Fig. 8-6) is used to switch the pickup roller on and off at desired times. Main logic generates a clutch control signal that is amplified by driver circuits before being fed to the electromagnet. When deactivated, the plunger disengages the

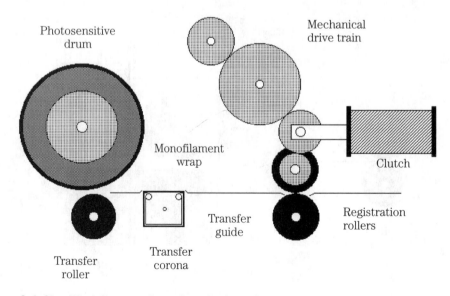

8-6 Simplified diagram of a registration/transfer area.

pickup roller from the drive train. When activated, the plunger engages the pickup roller, which causes the pickup roller assembly to turn and grab the next available piece of paper. A separation pad beneath the pickup roller prevents more than one sheet from being taken at any one time. Paper stops when it reaches the idle registration rollers. Notice how paper will bow—this is a normal and harmless function in the paper path.

If the main motor operates, but the pickup roller does not turn (you can see this during an open-cover test print), inspect the pickup roller clutch solenoid. Note there are probably two major solenoids—one for the pickup roller, and one for the registration rollers. When printing starts, one of the two solenoids (the pickup solenoid) should engage immediately. If no solenoid engages, there is an electronic problem. If the pickup solenoid engages, but your pickup roller does not turn (or does not turn properly), repair or replace the pickup mechanical assembly. There might be a faulty clutch or other mechanical defect.

When the pickup solenoid fails to actuate, use your multimeter to measure voltage across the solenoid. You should see voltage toggle on and off as the solenoid is switched. If voltage changes, but the solenoid does not function, replace the solenoid. If voltage remains at zero (or does not switch from some other voltage), there is probably a fault in the solenoid driver circuit. Troubleshoot the solenoid signal back into main logic. You also might simply replace the pickup clutch PC board. If the problem persists, replace the ECP.

Finally, you can check for feed roller wear by measuring the distance between the trailing edge of the paper and the end of the paper tray just as the sheet stops. This time occurs between the point where the pickup roller stops, and the registration rollers start. Normally, this trailing edge should advance about 3 inches or more. If it does not advance this far, your pickup roller is probably worn out. Replace the pickup roller assembly and separation pad.

Registration/transfer area Registration rollers hold on to the page until its leading edge is aligned with the drum image. Force is supplied by the main motor, but another electromagnetic clutch switches the registration rollers on and off at the appropriate time. Once paper and the drum image are properly aligned, main logic sends a clutch control signal that is amplified by driver circuits to operate the registration clutch solenoid. After the clutch is engaged, registration rollers will carry the page forward to receive the developed toner image. The registration/transfer assembly usually consists of registration rollers, the drive train, a registration clutch solenoid, a transfer guide, and the transfer corona assembly as shown in Fig. 8-6. If paper enters the printer but does not reach the fusing rollers, your fault is probably in this area.

You can see registration roller operation by opening a housing, defeating any corresponding interlocks, and defeating any EP cartridge sensitivity switches, then initiating a self-test. **Use extreme caution to prevent injury from high-voltage or optical radiation from the writing mechanism—especially from lasers.** Watch the paper path and drive train very carefully. If you see an obstruction in the paper path or in the drive train, unplug the printer and allow 10 minutes for the fusing assembly to cool before attempting to clear the obstruction! Replace any gears or bushings that appear damaged or worn. Pay close attention to any tension equalizing springs (called *torsion* springs) attached to the registration rollers. Reseat or replace any torsion springs that might be damaged or out of position.

Carefully inspect the monofilament line encircling the transfer corona. Make sure that the line is intact, and not interfering with your paper path. Do not approach the corona with your hands or any metal tool! If you see signs of physical interference, unplug the printer and allow high-voltage to discharge before replacing the monofilament line or transfer corona assembly.

If your main motor and drive train operate, but registration rollers do not turn (or turn properly), inspect the registration solenoid clutch. It will usually be adjacent to the pickup solenoid clutch. The solenoid should engage moments after the pickup roller disengages. If the registration solenoid does not engage, there is an electrical problem. If the solenoid does engage, but registration rollers do not turn, your mechanical clutch or registration rollers are probably worn out. Replace the mechanical registration assembly.

When your registration solenoid fails to actuate, measure the signal driving the solenoid. You should see the signal toggle on and off with the solenoid. If the signal changes, but the solenoid does not fire, replace the jammed or defective solenoid. If voltage does not change, there is probably a fault in the solenoid driver circuit or main logic. Troubleshoot the solenoid signal back into your main logic. You might replace the registration clutch PC board. If the problem persists, replace the ECP.

Exit area At the exit area, a page has been completely developed with a toner powder image. The page must now be compressed between a set of fusing rollers—one provides heat, and the other applies pressure. Heat melts the toner powder, and roller pressure forces molten toner permanently into the paper fibers to fix the image. As a fixed page leaves the rollers, it might stick to the fusion roller. A set of evenly spaced separation pawls pry away the finished page, which is delivered to the output tray. This step completes the paper feed process as shown in Fig. 8-7. Main motor force is delivered to the fusion rollers by a geared drive train. There are no clutches involved in exit area operations, so the drive train moves throughout the entire printing cycle.

It is important that the printer detect when paper enters and leaves the exit area. Based on paper size and fusion roller speed, a page has only a set amount of

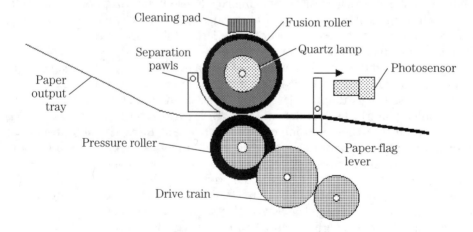

8-7 Simplified diagram of the fusing/exit area.

time to enter and leave the exit area before a PAPER JAM is initiated. To detect the flow of paper, an optoisolator is usually actuated by a weighted plastic lever. An example lever assembly is shown in Fig. 8-7.

Normally, a paper-flag lever protrudes down through a slot in the empty paper path, which leaves the optoisolator clear. Its resulting logic output indicates no paper. When paper reaches the lever, it is pushed up to the paper level. This action, in turn, moves the flag into the optoisolator slot, causing a logic change that shows paper is present—a timer is started in main logic. When everything works properly, paper moves through the fusion roller assembly. As paper passes, the lever falls again, returning the optoisolator to its original logic state. If the optoisolator returns to its initial value before the timer expires, it means that paper has moved through the exit area properly. If paper remains, a PAPER JAM is indicated.

A long-term timer was started at the beginning of the printing cycle. If this long-term timer expires before paper reaches the paper flag lever, a PAPER JAM also is generated. As you might suspect, there are a variety of problems that can cause a jam error.

Begin by checking the paper path for any obstructions. Unplug the printer and wait 10 minutes for the assembly to cool before exposing the fusing assembly. It might be necessary to remove secondary safety guards covering the heater roller. Remove any obstructions or debris that you find to be interfering with the paper. Make sure that your plastic separation pawls are correctly attached. Clean the pawls if they appear dirty. Inspect the paper flag lever carefully to be sure that it moves freely. Replace the flag lever assembly if it appears damaged or worn-out. Also check all interconnecting cables and wiring to see that the paper lever optoisolator is attached.

The drive gears that run your fusion rollers are often attached to a door housing. In this way, fusing rollers are disengaged whenever that access door is opened. This set of gears is sometimes called the *delivery coupling* assembly. If these gears are not engaging properly because of wear or damage, the fusion rollers will not operate (or operate only intermittently). Repair or replace any faulty delivery coupling components.

If the mechanics of your exit area appear to be operating correctly, you should examine the operation of the paper flag optoisolator. Replace any safety guards for the fusing assembly. Turn on the printer and use your multimeter to measure voltage across the optoisolator output. Note that you might have to defeat any open cover interlocks to ensure proper voltage in the printer. **Use extreme caution when measuring, and stay well clear of the high-voltage coronas.** Move the paper lever to actuate the optoisolator. You should see the output voltage toggle on and off as the optoisolator is actuated. If output voltage does not change, replace the faulty optoisolator. If voltage changes as expected, but paper jams are still indicated, troubleshoot your sensor signal back into the main logic circuit. If the problem persists, replace the ECP.

Symptom 3 *The printed image appears with a smudged band and overprint (Fig. 8-8).* This type of symptom is usually the result of a worn or damaged paper-pickup assembly. A worn pickup assembly can allow the pickup roller to rotate past its idle position such that it is pressing slightly on the page. If this happens while the registration rollers try to transfer the page into the printer, friction can delay paper movement for a moment. This delay results in a dark band that appears rather like

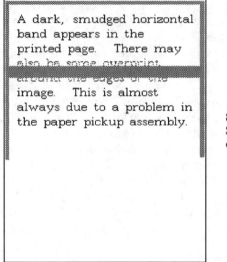

A dark, smudged horizontal band appears in the printed page. There may also be some overprint around the edges of the image. This is almost always due to a problem in the paper pickup assembly.

8-8
Smudged horizontal band and overprint.

an overprinted smudge. The solution to this problem is to replace the entire pickup assembly. You also might wish to replace the pickup roller clutch. Check the associated gear train to make sure there are no obstructions or damaged components.

Sensor and interlock problems

Sensors play a major role in EP printer operation. With so many variables in the outside world influencing the ultimate print quality that you see in a finished page, it is vital that an EP printer detect the physical conditions within the printer. For the purposes of this book, the primary interest is in monitoring temperature, as well as the state of physical contacts. There are three types of sensors used for these purposes: resistive sensors, mechanical sensors, and optical sensors.

Resistive sensors

Electrostatic printers use a fusion-roller assembly to apply heat and pressure to fix a toner image on paper. Fusion temperature must be carefully maintained at about 180°C to achieve an optimum toner melt. To control temperature, it is necessary to detect temperature, which is the job of a *thermistor*. *Thermistors* (thermal resistors) are resistors whose values change in proportion to their temperature. Depending on their formulation, thermistors can be constructed to increase or decrease with temperature. For the purpose of this book, a thermistor is assumed to increase its value with temperature.

A simple temperature alarm is shown in Fig. 8-9. The adjustable resistor (R_{adj}) sets the alarm trip point, and the thermistor (T) forms a voltage divider with the fixed resistor (R). While temperature is below the alarm setting, V_{sens} exceeds V_{ref}, so the output is about +V as shown. When temperature rises above the alarm setting, V_{sens} drops below V_{ref}, so the output drops to about ground (0 V).

Because a thermistor value is roughly proportional to temperature, it can be used in a proportioning circuit to regulate such things as fusion roller temperature.

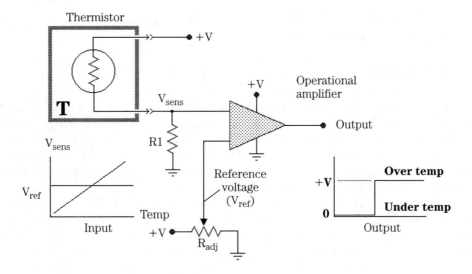

8-9 Schematic of a simple temperature-alarm circuit.

A proportioning circuit is shown in Fig. 8-10. As temperature increases, the oscillator will produce shorter pulses, and vice versa. The output can be detected by a microprocessor or ASIC, or used to drive a heater control circuit directly.

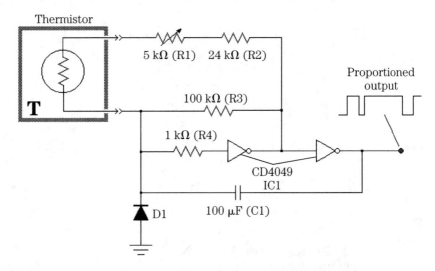

8-10 Schematic of a simple temperature-proportioning circuit.

Mechanical sensors

When position or presence must be detected, a simple mechanical switch can be used. A set of mechanical contacts might be normally open or closed, then actuated by the presence of paper, the paper tray, closed housings, etc. The condition of each switch (whether used individually or in sets) is often detected and acted upon directly by a logic circuit. The switches used in a printer's control panel often are considered to be mechanical sensors because they are detecting your input from the outside world.

Optical sensors

Mechanical sensors generally lack reliability over long-term use. Electrical contacts wear out through use and environmental corrosion. Contacts also are subject to electrical *ring*—an output that might vary on and off for several milliseconds before reaching a stable condition. Optical sensors are immune to these problems.

A basic optical sensor (called an *optoisolator*) is shown in Fig. 8-11. An optoisolator consist of two parts: (1) a transmitter, and (2) a receiver. Both are separated by a physical gap. The transmitter is usually an *IR* (infrared) LED kept on at all times. The receiver is typically a photosensitive transistor that is most sensitive to light wavelengths generated by the LED. When its gap is clear, light passes through unobstructed and saturates the phototransistor, in turn producing a logic 0 output signal. If an object (such as paper or the carriage) interrupts the light path, the phototransistor will turn off and allow a logic 1 output signal. There are no moving parts in an optoisolator, so it can operate at high speeds, and it will never wear out mechanically. Your printer paper supply and paper exit sensors are typically optical.

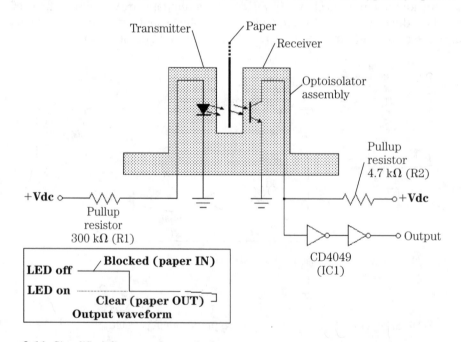

8-11 Simplified diagram of an optical sensor.

Troubleshooting sensors and interlocks

Before performing sensor checks, make it a point to examine any connectors or interconnecting wiring that tie the sensor into its conditioning circuit. Be certain that all cabling is installed correctly and completely before proceeding. Also remember that most sensors are mounted on small PC board assemblies that might contain other sensors or electronics. If you do not have the schematics or test instruments to check a sensor as discussed below, it is acceptable simply to replace that sensor PC board outright. If you do check the sensor and it checks properly (and its conditioning circuit appears to be functional), there is likely a problem with the ECP.

Symptom 1 *You see a PAPER OUT message even though paper is available. Also, the error might not appear when paper is exhausted.* If your paper sensor is a mechanical switch, place your multimeter across its leads and try actuating it by hand. You should see the voltage reading shift between a logic 1 and logic 0 as you trigger the switch. If you measure some voltage across the switch but it does not respond (or responds only intermittently) when actuated, replace the defective switch. If it responds as expected, check its contact with paper to be sure that it is actuated when paper is present. You might have to adjust the switch position or thread paper through again to achieve better contact.

Check an optical paper sensor by placing your multimeter across the photosensitive output; then, try to actuate the sensor by hand. You might need to place a piece of paper or cardboard in the gap between transmitter and receiver. You should see the phototransistor output shift between logic 1 and logic 0 as you trigger the optoisolator. If it does not respond, check for the presence of dust or debris that might block the light path. If excitation voltage is present, but the phototransistor does not respond, it is probably defective. Replace the optoisolator or sensor PC board assembly. When a sensor responds correctly, the trouble is probably in your ECP. If you do not have schematics or test instruments to troubleshoot in detail, simply replace the ECP.

Symptom 2 *Fusing temperature control is ineffective. Temperature never climbs, or climbs out of control. Temperature fluctuations affect print quality or initialization for EP printers.* Unplug the printer and allow at least 10 minutes for the printer to cool and discharge. Disconnect the thermistor at its connector. Use your multimeter to measure its resistance. A short or open circuit reading might indicate a faulty thermistor, so replace any suspect part. If you get some resistance reading, warm the thermistor with your fingers and see that the reading changes (even a little bit). A reading that does not change at all suggests a faulty thermistor. Never touch a hot thermistor with your fingers! You might have to replace the thermistor PC board.

If the thermistor is intact, the problem is likely in the ECP. Reconnect the thermistor, restore printer power, and use your oscilloscope to check the output of the thermistor conditioning circuit on the ECP. If you do not have schematics or test instruments to check the ECP in detail, replace the ECP outright.

Symptom 3 *You see a PRINTER OPEN message.* Printers can be opened to perform routine cleaning and EP cartridge replacement. The cover(s) that can be opened to access your printer are usually interlocked with the writing mechanism and high-voltage power supply to prevent possible injury from laser light, fusing heat, or high-voltages while the printer is opened. A typical interlock assembly is shown in Fig. 8-12. The top cover (or some other cover assembly) uses a simple pushrod to actuate an electrical switch. When the top cover is opened, the interlock switch opens, and the printer driver voltage (+24 Vdc is shown) is cut off from all other circuits. In some printers, high-voltage is cut off directly, which effectively disables printer operation. When the top cover is closed again, the interlock switch is reactivated, and printer operation is restored.

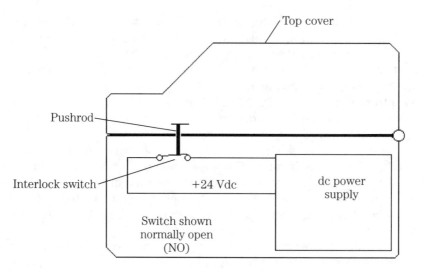

8-12 Simplified diagram of a cover interlock assembly.

First, make sure that your cover(s) are all shut securely (try opening and re-closing each cover firmly). Inspect any actuating levers or pushrods carefully. Replace any bent, broken, or missing mechanical levers. Unplug the printer and observe how each interlock is actuated (it might be necessary to disassemble other covers to observe interlock operation). Adjust the pushrods or switch positions if necessary to ensure firm contact.

Unplug the printer and use your multimeter to measure continuity across any questionable interlock switches. It might be necessary to remove at least one wire from the switch to prevent false readings. Actuate the switch by hand to be sure that it works properly. Replace any defective interlock switch, re-attach all connectors and interconnecting wiring, and re-test the printer.

If a switch itself works correctly, check the signals feeding the switch. Check the dc voltage at the switch. If the voltage is low or absent, trace the voltage back to the power supply or other signal source at the ECP. If signals are not behaving as expected, or PRINTER OPEN message remains, trace the interlock signal into the main logic board and troubleshoot your electronics. If you do not have schematics or test instruments, you should replace the ECP outright.

Scanner-motor/main-motor problems

Delivering an accurate, well-timed, modulated laser beam to a drum surface is no easy feat. Laser printers have evolved over the last decade from large, heavy, clunky (and delicate) mechanisms with discrete lasers, mirrors, and optics, into svelte, rugged workhorses that integrate the laser, scanner, and some optics and electronics right into a single, easily replaceable laser/scanner assembly. You learned how to deal with laser delivery problems in chapter 7. Here, you will see how to deal with scanner and main motor problems. Keep in mind that a laser/scanner assembly is *only* needed for ES printers using laser writing mechanisms. LED or LCS writing mechanisms do not use scanning mirrors.

Symptom 1 *You see a general SCANNER ERROR message.* The scanner is an optical-grade hexagonal mirror driven by a small, brushless dc motor that operates independently of the main motor. Printing will only be enabled after the scanner has reached its proper operating speed. The scanner is engaged at the beginning of a printing cycle. For many laser printers, you will recognize the scanner motor by a somewhat distinctive sound. Hewlett-Packard calls it a *variable pitch whirring noise.* Motor speed is constantly monitored and controlled by main logic. If the motor fails to turn when power is applied, or at any point during the printing process, a SCANNER ERROR is generated. There are two sources of problems here. Either the laser/scanner assembly has failed, or the ECP motor control circuit has developed a fault.

Unplug the printer, open its housings, and carefully inspect connectors and interconnecting wiring between the laser/scanning unit and main logic circuits. Reseat any connectors or wiring that appears to be loose. If the problem seems intermittent, try a new cable assembly. The scanner is usually tested briefly during printer initialization. If you cannot hear the scanner motor, use your multimeter to measure the dc excitation voltage across the motor. A correct voltage reading suggests a defective scanner motor, so replace the entire laser/scanning assembly. A low or missing excitation voltage indicates a defect in your main logic or driving circuitry that switches its motor voltage on and off. Troubleshoot the excitation voltage and switching circuitry back into the main logic board. Remember that you might have to defeat cover interlocks to enable the printer low-voltage power supply. If your readings are inconclusive or you can't troubleshoot the main logic circuits, replace the ECP.

Symptom 2 *The main motor does not turn, or turns intermittently.* The main motor is responsible for providing all of the mechanical force that drives an EP printer. If the main motor fails to work, or works only intermittently, all printer operation will cease. Open the printer and defeat the cover interlock(s) as required to allow the printer to work. **Use extreme caution to prevent shocks or burns during printer operation.** Make sure that the motor's cables are attached completely and correctly.

If the motor is properly connected, check motor operation during a self-test. When the motor turns but the gears and other mechanics do not, there is a problem with the mechanical assemblies. Something is loose, jammed, or damaged. Find and clear any jam that you might find. If you locate a damaged component, replace the defective mechanical assembly. If the motor itself does not turn, check each output from the dc power supply. A low supply voltage can result in motor problems. If you find a low or absent output, replace the dc power supply. You also might troubleshoot the supply as in chapter 6. If all supply voltages appear correct, replace the main motor assembly outright. Use care when reassembling the mechanical components. If the trouble persists, the fault is in the ECP. Try replacing the ECP.

EP cartridge problems

The electrophotographic (EP) cartridge plays an important role in the operation of your printer. An EP cartridge contains the photosensitive drum, primary corona, development roller, and toner supply. By placing these vital components into an easily replaceable assembly, it is a simple matter to maintain the printer—all the major components that develop the image are exchanged every time the EP cartridge is replaced.

One attribute of the photosensitive drum is that not all drums have the same sensitivity. When a drum is manufactured, its sensitivity is tested, and tabs are placed on the EP cartridge. When the cartridge is installed, its tabs actuate a set of microswitches in the printer. The ECP reads the settings of these microswitches to set laser power. Lower-sensitivity cartridges use higher laser power, and higher-sensitivity cartridges use lower laser power.

Symptom 1 *You see a NO EP CARTRIDGE message.* An electrophotographic (EP) cartridge assembly uses several tabs (known as *sensitivity tabs*) to register its presence, as well as inform the printer about the relative sensitivity level of the drum. Main logic regulates the output power of its writing mechanism based on these tab arrangements (that is, high-power, medium-power, low-power, or no-power—NO CARTRIDGE). Sensitivity tabs are used to actuate microswitches located on a secondary PC board. The sequence of switch contacts forms a code that is interpreted by main logic.

Begin by checking the installation of your current EP cartridge. Make sure that it is in place and seated properly. Check to be sure that at least one tab is actuating a sensor switch. If there are no tabs on the EP cartridge, replace it with a new or correct-model EP cartridge having at least one tab. Re-test the printer. If your NO EP CARTRIDGE error persists, check all sensitivity switches.

Unplug the printer and use your multimeter to measure continuity across each sensitivity switch. It might be necessary to remove at least one wire from each switch to prevent false continuity readings. Actuate each switch by hand and see that each one works properly. Replace any microswitch that appears defective or intermittent. Replace any connectors or interconnecting wiring, and re-test the printer.

If you still receive an error message, troubleshoot each switch signal into the main logic board. There might be a problem with your signal conditioning circuits or main logic components. You might simply replace the ECP outright.

Symptom 2 *You see a TONER LOW message constantly, or the error never appears.* A toner sensor is located within the EP cartridge itself. Functionally, the sensor is little more than an antenna receiving a signal from the high-voltage ac developer bias as shown in Fig. 8-13. When toner is plentiful, much of the electromagnetic field generated by the presence of high-voltage ac is blocked. As a result, the toner sensor only generates a small voltage. This weak signal is often conditioned by an amplifier using some type of operational amplifier circuit that compares sensed voltage to a preset reference voltage. For the comparator of Fig. 8-13, sensed voltage is normally below the reference voltage, its output is a logic 0. Main logic would interpret this signal as a satisfactory toner supply. As toner volume decreases, more high-voltage energy is picked up by the toner sensor, in turn developing a higher voltage signal. When toner is too low, sensed voltage will exceed the reference, and the comparator's output will switch to a logic 1. This operation is handled in main logic, and a TONER LOW warning is produced.

Unfortunately, there is no good way to test the toner sensor. High voltage is very dangerous to measure directly without the appropriate test probes, and the signals picked up at the receiving wire are too small to measure without a sensitive meter or oscilloscope. Begin your check by shaking the toner to redistribute the toner supply (or insert a fresh EP cartridge). Refer to the user's manual for your particular printer

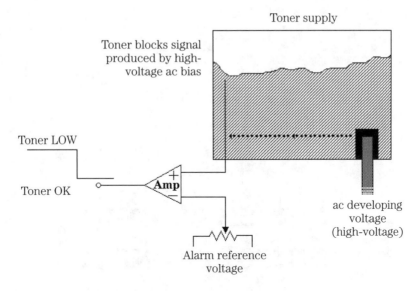

8-13 Simplified schematic of a toner sensor.

to find the recommended procedure for redistributing toner, then re-test the printer. If the error message persists, try replacing the EP cartridge.

Your next option is to repair or replace the high-voltage power supply. If you have the proper test instruments to measure high-voltage safely, check the high-voltage level at your developing-bias connector. If this voltage is absent or low, repair or replace the high-voltage supply. **Use extreme caution when attempting a high-voltage repair! Allow plenty of time for the supply to discharge before disassembling the printer.** If the power supply appears defective, you should re-place the high-voltage supply outright. If your high-voltage system checks correctly, there is probably a fault in the ECP. Replace the ECP.

9
CHAPTER

The electronic control package

Regardless of how simple or how sophisticated your electrophotographic printer might be (Fig. 9-1), every operation is controlled by a set of electronic circuits that is commonly called the *controller* or *ECP* (electronic control package). The specific architecture and components used in ECPs varies between printer models and manufacturers, but the basic set of operations to be performed are remarkably similar.

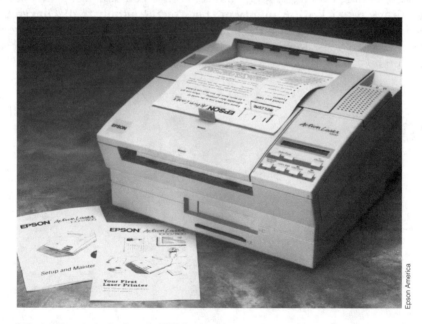

9-1 An Epson ActionLaser 1500 laser printer.

An ECP must communicate with the outside world (the user as well as the computer), and translate communicated data—often in the ASCII format—into characters or patterns of dots that are used to form an image. The ECP also directs such physical tasks as paper pickup and high-voltage control. It interprets sensor information regarding paper supply, EP cartridge condition, and fusing temperature. Finally, each of these tasks (and more) must be coordinated to work together.

A typical ECP can be broken down into four functional areas: (1) communication (or *interface*), (2) memory, (3) control panel, and (4) main logic. Figure 9-2 is a block diagram for a typical EP printer. Note that these functions often can be fabricated onto a single PC board. Before you begin troubleshooting an ECP, you should thoroughly understand the operations and components in each of these sections.

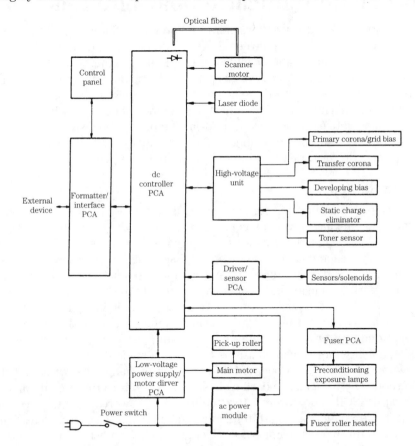

9-2 Diagram of a typical laser printer. Hewlett-Packard Co.

Communication

Your printer must communicate with the outside world to receive the characters or graphics data that it must print. Binary data representing this information is sent from the host computer to your printer more than one of several possible communication links. The computer also receives commands and status information back from the printer. These return signals are used to regulate the flow of data. Although there are many variations of communication links, data is transferred using either a *serial* or *parallel* technique. Data is sent over a parallel link as whole characters—that is, all the binary digits (or bits) that compose a character are sent at the same time over multiple signal wires. A serial link transfers data one bit at a time over a single wire. Before you see the specifics of parallel or serial communication, some background information is in order. Experienced troubleshooters can skip the next few sections.

Printer communication background

A computer can send three types of information to a printer: (1) text characters, (2) control codes, and (3) graphics data. Keep in mind that not all printers will accept character or graphic data, or interpret them, in the same way. Character data represents text—letters (in any language), numbers, punctuation, or other text symbols. Control codes are used to send commands to the printer. Control codes can set general operating modes such as font style, enhancements, or pitch, but they also can direct immediate operations such as form feed or line feed. Using control codes eliminates the need to operate a control panel manually while a document prints. Other control codes provide the printer with graphic data.

ASCII explained

Before any communication can take place, both the printer and computer must speak the same language—when a computer sends out the character *H*, its printer must recognize that character as an *H*. Otherwise, it will just print unintelligible garbage. Because each character and control instruction is represented by its own unique numerical code, both printers and computers must use a common set of codes that describe some minimum number of codes. In the early days of computers, each manufacturer had its own code set. You can probably imagine how difficult it was to combine equipment made by different manufacturers. As the electronics industry matured and printers became more commonplace, the demand for equipment compatibility forced manufacturers to accept a standard character code set.

The American Standard Code for Information Interchange (known as ASCII) has come to represent a single, standard code set for computer/printer communication. The standard ASCII code covers letters (upper- and lowercase), numbers, simple symbols for punctuation and math functions, and a few basic control codes. For example, if you want to print an uppercase *D*, your computer must send the number 68 to the printer. The printer would then translate 68 into a dot pattern that reflects the selected font, character pitch, and enhancements to form the letter *D*. To print the word "Hello," a computer must send a series of numbers: 72, 101, 108, 108, and 111. Pure (original) ASCII uses codes 0 to 127.

Because of the way character codes are actually sent, however, most computers also can use codes ranging from 128 to 255, but keep in mind that any code over 127 is not pure ASCII. Instead, codes from 127 to 255 are sometimes called an *alternate character set.* Such an alternate character set can contain single-block graphic characters, Greek symbols, or other language characters. In some cases, codes 128 to 255 just duplicate codes 0 to 127. If your computer sends a code from 128 to 255, you might be printing characters that are different from those on your computer screen.

Control codes

Not only must a computer specify what to print, it also must specify how to print. Form feeds, font styles, and enhancements are some of the controls that a computer must exercise to automate the printing process. Just imagine the confusion if you stand by your printer to change these modes manually while a document was printing! Unfortunately, control codes are often the cause of some incompatibilities between computers and printers.

When ASCII was first developed, printers were extremely primitive by today's standards. Multiple fonts and type sizes, graphics, and letter-quality print had not even been considered. Few controls were needed to operate these early printers, so only those few critical controls were incorporated into ASCII. You might recognize such controls as form feed (FF), line feed (LF), or carriage return (CR).

With the inclusion of advanced electronic circuitry, a greater amount of intelligence became available in printers (especially EP printers). This intelligence has made so many current printer features possible. ASCII codes are still standard, but there simply are not enough unused codes to handle the wide variety of commands that are needed. Manufacturers faced the choice of replacing ASCII (and obsoleting a long-established and growing customer base), or developing a new scheme to deal with advanced control functions. Ultimately, manufacturers responded to this by devising a series of multi-code control sequences. These were known as *escape sequences* because the ASCII code 27 (escape) is used as a prefix. EP printers make extensive use of escape sequences.

Printer capabilities can vary greatly between models and manufacturers. As a result, escape codes are not standard. If computer software is not written or configured properly for its particular printer, control codes sent by the computer might cause erratic or unwanted printer operation.

Escape sequences are typically two or three ASCII codes long, and each begins with ASCII code 27. The escape character tells the printer to accept subsequent characters as part of a control code. For example, to set a laser printer to SELF-TEST, a computer might have to send an ASCII code 27, followed by an ASCII code 122 (*z*). Software in the printer main logic would interpret this code sequence and alter the appropriate modes of operation accordingly. Multi-code sequences will certainly become more common as printers get even more sophisticated.

Number systems

Not only must a computer and printer exchange codes that they understand, but every code must be sent using a number system that is compatible with electronic (digital) circuitry. You already know the decimal (or base 10) number system. The symbols 0 through 9 are used in combinations that can express any quantity. The symbols themselves are irrelevant—ten other symbols could as easily have been used, but 0 through 9 are the ones accepted down through the centuries. What is important is the quantity of characters in a number system. In decimal, one character can express 10 unique levels or magnitudes (0–9). When the magnitude to be expressed exceeds the capacity of a single character, the number carries over into a higher representative place, which is equal to the base of the system raised to the power of that place. For example, the number 276 has a 2 in the hundreds place, a 7 in the tens place, and a 6 in the ones place. You have worked with this since grammar school.

If electronic circuits could recognize ten different levels for a single digit, then digital electronics would be directly compatible with the human decimal system, and ASCII codes would be exchanged directly in their decimal form. However, digital electronics can only recognize two signal levels. These conditions are ON or OFF (true or false). This system is the *binary* (or base 2) number system. Because only two conditions can be expressed, only two symbols are needed to represent them. The symbols 0 (OFF) and 1 (ON) have come to represent the two possible conditions for a binary digit (commonly called a *bit*). ASCII codes must be sent as sets of binary digits.

As with the decimal number system, when a quantity to be expressed exceeds the capacity of a character (here the character is a 1) the number carries over into a higher place, which is equal to the base of the system (base 2) raised to the power of the place. You have probably seen binary signals expressed as 2^n, where n is the bit place position. As Fig. 9-3 shows, the decimal number 20 equals the binary number 10110. A lower-case u with an ASCII code of 117 would be expressed as 1110101 in a digital system. Keep in mind that seven bits can express numbers from 0 to 127. Eight bits can express numbers up to 255.

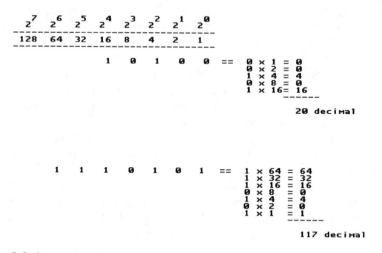

9-3 Converting binary numbers into decimal numbers.

Binary digits

To have any meaning at all in electronic circuits, there must be a clearly defined relationship between a binary digit and a voltage level. Because a binary 1 is considered to be an ON condition, it usually indicates the presence of a voltage. A binary 0 is considered OFF, so it denotes the absence of a voltage. In reality, the actual amount of voltage that describes a 1 or 0 depends upon the logic family in use. Common digital circuits using conventional *TTL* (transistor-transistor logic) ICs classify a logic 0 as 0.0 to +0.8 Vdc, and a logic 1 as +2.4 to +V_{cc} (the voltage powering the IC).

Communication links

Information, in the form of binary digits, must traverse the physical distance between a computer and printer by a communication link—a wire cable. The construction and characteristics of this cable will depend upon which method of communication is in use. There are two dominant methods of sending printer information: serial and parallel. A *parallel* link is easiest to understand because of its straightforward operation. Notice that eight bits of an ASCII code are transferred simultaneously (D0 through D7). Data lines alone, however, are not enough to transfer information successfully. Both the computer and printer must be synchronized so that the printer will accept data when it is offered, or ask the computer to wait until it is ready. Synchronization of a parallel link is accomplished using several control wires, in addition to data lines. Some control lines signal the printer, and others will signal the computer. This mutual coordination is known as *handshaking*.

Parallel operation is reasonably fast. The printer will accept information as fast as the computer can send it, often operating at speeds exceeding 1,000 *CPS* (characters per second). At eight bits per character, that amounts to more than 8,000 bits per second. The main disadvantage to parallel links is its limited cable distance. With so many high-speed data signals running together in the same cable, its effective length is a few meters. Beyond that, electrical noise and losses can cause distortion and loss of parallel data.

A *serial* link might appear simpler because of the simplified wiring requirement, but its actual operation is somewhat more involved. Two wires are used to transfer information. One of these wires carries data from computer to printer, and the other carries data from printer to computer. Because data can travel in both directions, this is known as a *bidirectional* data link. Only one wire is available to send (or receive), so a character must be sent one bit at a time. Serial data also must be synchronized between the computer and printer. To accomplish this over a single wire, synchronization bits are added at the beginning and end of each character. An extra bit (known as a *parity bit*) also might be included to allow error checking.

Serial handshaking can be provided either through hardware or software. Software handshaking takes advantage of the bidirectional nature of serial communication by allowing the printer to transmit control codes back to the computer. Two codes used commonly for software handshaking are XON and XOFF. Older serial handshaking might use the codes ETX and ACK, but EP printers rarely use that convention any more.

Hardware handshaking does not support data transfer from printer to computer. Instead, a serial handshake line signals the computer that the printer is busy. Some interfaces carry more than one handshaking line. Usually, you can expect to see a variety of handshaking schemes between printer generations, so pay particular attention to the wiring in your serial printer cable. In spite of their added operating complexity, serial communication is extremely popular because of its bidirectional nature, flexibility, and its ability to work well over long distances.

Communication standards

There are literally hundreds of ways that you can implement a communication interface. You can believe that a great many versions have been tried and abandoned since the early days of commercial printers. The evolution of technology favors the best methods and techniques, so those that work well and grow with advances in technology can sometimes develop into standards that other manufacturers adopt in the future. Standards are basically a detailed set of rules and performance characteristics that clearly define the construction, connection, and operation of a circuit or system—in this case, a communication interface. By adopting an established standard, manufacturers can be sure that printer brand Y will operate fine with computer brand X, and vice versa.

Parallel communication

A parallel communication link requires two sets of signal wires as shown in Fig. 9-4. One data line is needed for each bit of parallel data (usually eight lines). A ground wire is often supplied as a separate return path for each data line. The flow of data also must be coordinated with the computer. The coordination is accomplished through a series of *handshaking* lines. Both handshaking and data wiring is routed to a female connector mounted on the printer. This connector is attached to the computer over a multiconductor cable. The diagram of Fig. 9-5 shows parallel interface hardware in more detail.

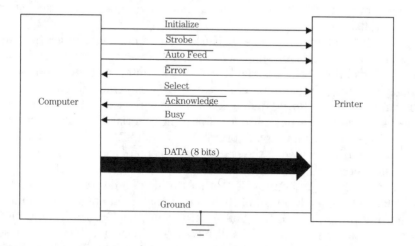

9-4 Data and handshaking lines of a parallel port.

When the computer sends out a character, it places all eight bits of the character on data line D0 through D7. Data from the computer follows the standard conventions established for TTL devices. A logic 0 is represented by a voltage level between 0.0 and +0.8 Vdc, and a logic 1 is represented by a voltage between +2.6 and +4.9 Vdc. You can easily measure these voltages with a logic probe or oscilloscope. The timing diagram in Fig. 9-6 provides a good illustration of parallel port operation.

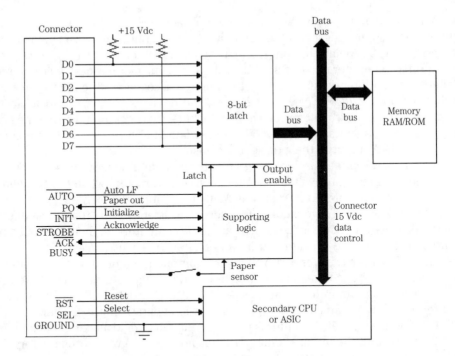

9-5 Simplified diagram of a parallel communication circuit.

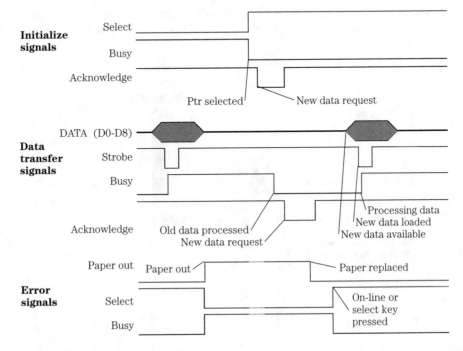

9-6 Parallel-port timing signals.

When all data bits are available at the interface, a Strobe Signal from the computer causes supporting logic to generate a pulse. This pulse latches all data bits into the printer. At this point, the computer can send another character, or go on to other work. After the printer receives a valid strobe signal, its microprocessor circuit reads data onto the printer data bus where it is stored in a segment of temporary memory called the *data buffer*. For an EP printer, the data buffer can be up to several megabytes depending on how much memory is fitted to the printer. During the printing process, characters are removed from the data buffer, processed by the microprocessor, then output to the writing mechanism.

Although the printer is accepting and storing data, it sets its Busy Signal to a logic 1 level. A computer will not strobe any more characters while the busy line is logic 1. After data is stored successfully in its data buffer, the busy signal will be released to a logic 0, and its Acknowledge Signal will strobe briefly to request another character. Once an acknowledge is generated, the computer can then send out a new character. This cycle is repeated until every character is sent successfully.

Realize that the transfer of data between computer and printer occurs independently of its interpretation and printing. In practice, communication can take place much faster than any printer can print—that is why a data buffer is provided. A printer can accept characters in large groups, then operate on those characters until the buffer is empty. If the buffer is not large enough to hold all data, the computer will stand by until the buffer is empty, then send another group of characters. The timing of data transfer is controlled entirely by handshaking lines.

Serial communication

A serial-communication link exhibits slightly different characteristics. In the serial circuit of Fig. 9-7, there are only two data-carrying lines: Transmit Data (Tx) and

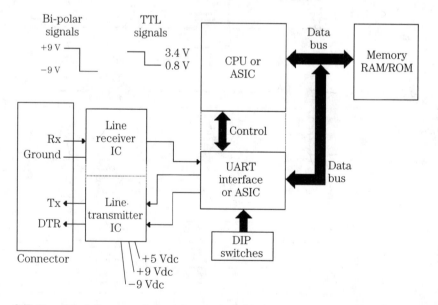

9-7 Simplified diagram of a serial communication circuit.

Receive Data (Rx). Any other remaining signal lines are for handshaking purposes. Notice that a serial interface is bidirectional. Data sent from the computer is received by the printer Rx line, while data (if any) sent back from the printer leaves its Tx line. Because a serial interface allows a printer to talk back to the computer, handshaking can be accomplished by transferring control characters back and forth. The serial communication bus is usually called RS-232. All signal and data lines are routed to a female connector mounted to the printer. Figure 9-8 shows the pinout of a serial communication port.

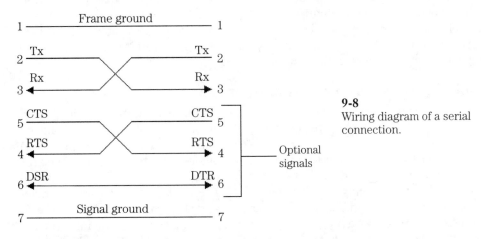

9-8
Wiring diagram of a serial connection.

Notice that the signal levels of serial data and control lines also are different from those of a parallel interface. Serial signals are bipolar—that is, one logic level is represented by a positive voltage, and the opposing level is represented by a negative voltage. This kind of bipolar operation allows serial interfaces to carry data over greater distances with less noise than parallel interfaces. Although Fig. 9-7 shows bipolar (loosely referred to as *analog*) signals ranging from + 9 to –9 V, a serial interface can use voltage levels from ±5 to ±15 V. You can measure these serial levels with an oscilloscope.

Unfortunately, bipolar voltages are not comparable with the digital logic devices at work in the printer, so data and handshaking signals must be translated between bipolar and TTL levels as required. This transition is accomplished by a set of devices known as *line transceivers*. For example, bipolar bits received from a computer are converted to TTL levels using a *line receiver*, and TTL bits must be converted to bipolar levels using a *line driver*.

The problem with serial data is that it can only be transferred one bit at a time. Each bit must be assembled into a complete character before it can be stored in memory. Just the reverse is true for control codes sent back to the printer. A character (usually eight bits long) must be disassembled and sent as a series of individual bits. This task of data manipulation is handled by a specialized IC that can perform these conversions. The basic name for such a device is a UART (universal asynchronous receiver/transmitter). Not only does a UART perform data conversions, but it also can attach or remove the overhead bits of a serial character. Overhead bits include such things as the Start Bit, Stop Bit(s), and Parity Bit.

When a bipolar bit is received by the printer on its Rx line, a line receiver translates it into a corresponding TTL level. Each bit enters the UART where overhead bits are stripped away. After an entire character has been assembled, the microprocessor circuit accepts that character onto the printer data bus, where it is stored in the data buffer to await further processing.

Most serial printer interfaces operate in either a hardware or software handshaking mode, and communication can be coordinated with discrete signal lines (such as DTR in Fig. 9-7) or control characters sent back from the printer. A character to be sent must be loaded into the UART. The microprocessor determines which character to send, then writes that character over the data bus. Once loaded, the character is disassembled, overhead bits are added, and each TTL bit is sent through a line driver IC. The line driver translates each TTL bit into a corresponding bipolar bit prior to leaving the printer.

Isolating the communication interface

A communication interface involves much more than printer circuits. The successful transfer of data requires proper operation of a computer and interconnecting cable as well. Trouble in any one of these three areas can interrupt the flow of data. Before you disassemble your EP printer, you should isolate the problem to the printer itself. The quickest and most certain way to do this is to test a working printer (one that you know is working well) on your existing computer using the same parallel or serial interface. If a working printer works properly, then you have ruled out the computer, cable, or software program. If a working printer also fails to operate, you might have a problem in your computer, its software configuration, or the cable.

Once you have isolated the problem to your printer, run a printer self-test. The self-test will test the printer motors, memory, writing mechanism, power supply, and most of the ECP. If a self-test pattern looks good, you can be pretty certain that the printer interface circuit is defective. If the self-test pattern is faulty, then your printer is suffering from a defect elsewhere.

Troubleshooting a parallel interface

Symptom *Printer does not print at all. There might be a PRINTER NOT READY error displayed on the control panel. The printer self-test looks correct.* Begin by examining your interface cable. If it is loose at either end, data and handshaking signals might not reach the printer. If you have run another printer successfully using your current cable, then the cable is almost certainly good. If you are in doubt, try a new cable. If you wish to check the original cable, disconnect the cable, and use your multimeter to measure continuity across each wire. Wiggle the cable to stimulate any possible intermittent connections. Replace any defective interface cable. If you do not have the tools or inclination to perform detailed procedures, replace the printer's ECP outright.

Disassemble your printer and expose its communication circuitry. With the printer connected and running, use your logic probe or oscilloscope (as shown in Fig. 9-9) to examine each handshaking line. Table 9-1 is a listing of typical pin assignments, but refer to your user's manual for the specific pin designations used with your parallel interface. Connect your test instrument to the signal ground (SG) line, then measure each of the handshaking signals in turn.

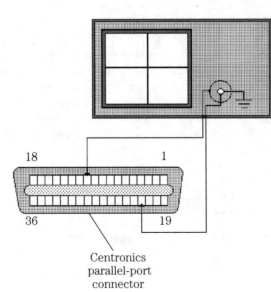

9-9
Testing a parallel-port connector.

Centronics
parallel-port
connector

**Table 9-1. Pin assignments
for a typical parallel interface**

Pin number	Signal	Name
1	Strobe	$\overline{\text{STR}}$
2	Data 0	D0
3	Data 1	D1
4	Data 2	D2
5	Data 3	D3
6	Data 4	D4
7	Data 5	D5
8	Data 6	D6
9	Data 7	D7
10	Acknowledge	$\overline{\text{ACK}}$
11	Busy	BUSY
12	Paper out	PO
13	Select	SEL
14	Auto line feed	$\overline{\text{AUTO}}$
15	—	—
16, 33	Signal ground	SG
17	Frame ground	FG
18	+5 Vdc	—
19–29	Data line grounds	—
30	Prime ground	—
31	Initialize	$\overline{\text{PRIME}}$
32	Error	$\overline{\text{ERROR}}$
34–36	—	—

There are four key status signals that communicate the current operating conditions of your printer: (1) Busy, (2) Select, (3) Paper Out, and (4) Error. Table 9-2 shows the interaction of these status lines versus the printer on-line or off-line condition. If you find these four signals at their appropriate levels when the printer is on-line, the printer should be ready to accept data. If on-line conditions are incorrect, a problem exists in your interface or main-logic circuitry. Check the supporting logic that provides your handshaking signals, or replace the ASIC or microprocessor that directs handshaking signals. If you do not have schematics or instruments to trace these signals, you might simply replace the ECP entirely.

Table 9-2.
Handshaking conditions
in a parallel interface

Condition	BUSY	SEL	PO	ERROR
Off-line	0	1	0	1
On-line	1	0	0	0
Paper out	1	0	1	0

Place your test instrument on any of the data lines (D0 through D7) and try printing under computer control. If your printer is correctly on-line, characters should begin arriving in rapid succession. This transmission will appear as a pulse signal on a logic probe, or as a random square wave on an oscilloscope. Keep in mind that you should see some indication of data as long as handshaking signals are set up for on-line status.

If you do not find a regular flow of data, watch the Acknowledge signal. It normally rests at logic 1, but it will pulse low briefly whenever the Busy line is returned to logic 0. It should appear as a pulse signal on a logic probe, or as a random square wave on your oscilloscope. An absent Acknowledge signal suggests a problem in supporting logic that handles handshaking. Check your supporting logic, or replace the ASIC or microprocessor that directs handshaking operation. If no technical data or instruments are available, you can replace the ECP.

Check parallel data into its latch shown in Fig. 9-4. As data enters the printer, you should be able to monitor the presence of any data bit. An active data line should appear as a pulse signal on your logic probe, or a random square wave on an oscilloscope. Repeat your check at each latch input. Refer to your printer schematic or manufacturer's data for the latch IC to determine input and output pins.

Monitor the latching pulse and output enable signal. These signals control data capture and transfer to the data bus. A Latch pulse is generated whenever a Strobe is received. The pulse loads data into the latch, but does not allow it on the output. When the microprocessor or ASIC is ready to store the character in the data buffer, an Output Enable signal places latched data onto the data bus. During normal operation, both control signals should appear as pulse signals on a logic probe, or as a random square wave on your oscilloscope. If both of these signals are present, the latch might be defective. Replace the latch and retest the printer. If the problem persists, replace the printer data buffer memory ICs. If either or both control signals are missing, there is a problem in your supporting logic or ASIC. Check your supporting logic or replace the ASIC. If you do not have the schematics or test instruments to perform this type of troubleshooting, replace the ECP outright.

Troubleshooting a serial interface

Symptom *Printer does not operate at all. A PRINTER NOT READY error might occur at the printer. The printer self-test looks correct.* Your first step should be to check the communication interface cable. If it is loose at either the printer or computer end, data and handshaking signals might not be able to reach the printer. If you have run another printer successfully using your existing computer and cable, then the cable itself is almost certainly good. To verify cable wiring, try a new cable. If you wish to test the cable, use your multimeter and measure continuity across each cable wire. Be careful here—many serial cables flip (or reverse) the Tx and Rx lines between computer and printer, so you must consider that when making measurements.

Serial communication requires a fairly large number of parameters to specify the structure and speed of each serial character. Word length, stop bit(s), parity, baud rate, and handshaking method, are some of the more common options that can be selected when setting up a serial communication link. However, each option must be set exactly the same way at both the computer and printer. If not, a printer cannot interpret just where data starts and ends. The resulting confusion will cause an erratic jumble of unintelligible print (if it prints at all). Communication parameters are usually set by software, or a series of jumpers or dip switches within the printer. Check these settings against those listed in your user's manual. If you have run another printer, check its configuration and compare settings. If you do not have the tools or inclination to perform detailed procedures, replace the printer's ECP outright.

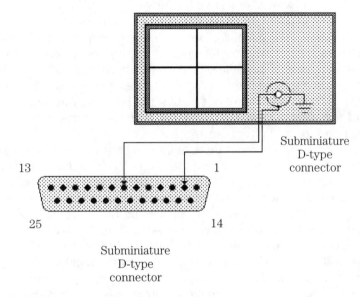

9-10 Testing a serial-port connector.

Disassemble your printer and expose its communication circuitry. With the computer running and connected, place your printer on-line, and use your oscilloscope (as shown in Fig. 9-10) to measure the activity on each handshaking line. Handshaking lines of primary interest to serial printers are DTR (Data Terminal Ready) and RTS (Request To Send). Note that RTS is simply ignored in many serial interfaces, so concern yourself primarily with DTR at the printer. DTR connects to DSR (Data Set Ready) at the printer. Figure 9-11 shows a typical serial link. Remember, if your printer is set up to use software handshaking, it might be unnecessary to inspect hardware handshaking lines.

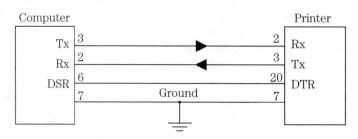

9-11 Diagram of a typical serial connection.

Serial signals at the interface connector are bipolar. If the printer is ready to receive data, its DTR line should be at a positive voltage (+5 to +15 V). A negative voltage (–5 to –15 V) indicates that the printer is not ready, so no data will be sent. If you have a schematic diagram for your printer, check the line driver circuit, and the signal from your UART (serial data functions can be integrated into an ASIC). You will have to refer to your printer schematic, or manufacturer's data for your specific ICs to determine which IC pins serve input and output functions. If you find a faulty handshaking line, replace the defective line driver or UART (or ASIC). If the serial port appears faulty, you also might replace the ECP outright

When your handshaking signal is correct (or your printer is configured for software handshaking), place your oscilloscope at the printer Rx line and try sending data from the computer. You should see characters begin to arrive immediately, so a random square wave varying from a positive to a negative voltage should appear on the oscilloscope. If this signal is absent, data is not reaching your printer. Check the conditions of any other handshaking lines that might be incorporated into your particular serial interface, and retest the printer.

Use your oscilloscope to measure data into and out of the line receiver IC. You should find bipolar data signals entering the line receiver, and a corresponding TTL (0 to 5 V) output from the receiver. Refer to your printer schematic or manufacturer's data for your particular line IC to find specific pin functions. If data does not leave the line receiver, replace the defective IC or replace the ECP entirely. TTL data signals also should enter the UART. If signals enter the UART (or ASIC) and the printer does not function, there might be a fault in the UART (or ASIC) or data buffer memory. Try replacing the UART (or ASIC), or replace the data buffer memory ICs. If you do not have schematics or technical data available for the printer, or you cannot test or troubleshoot the serial communication circuit, simply replace the ECP outright.

If your printer is operating in a software handshaking mode, then its DTR signal will usually be ignored. Control characters known as XON and XOFF are sent serially to the computer through the printer TX line. Older printers might use other codes. Under normal operation, the printer should send an XON character to the computer once an initialization has taken place. The computer can then begin sending data at will. When the printer data buffer is full, it sends an XOFF character that immediately stops all data transmission. An XON is sent along when the printer is ready to accept more data. Place your oscilloscope at the printer TX line and try sending data to the printer. You should find an occasional random square wave representing XON and XOFF handshaking codes as well as other data. This signal should be bipolar signal. If control codes are missing, handshaking data is not being sent to the computer. It is possible that the printer never sent a preliminary XON code when it was initialized.

If handshaking codes are absent, use your oscilloscope or logic probe to measure TTL data entering the line driver IC. Remember that the printer must be trying to print in order for any data to be sent. You should see a random TTL square wave representing XON and XOFF characters as well as other data. If this signal appears present, but there is no bipolar Tx output, replace your defective line driver IC. If there are no TTL signals reaching your line driver, your UART (or ASIC) is probably faulty. If you do not have the technical information or instruments to trace these signals, you might simply replace the ECP outright.

Memory

A laser or LED printer accepts data and control codes from its host computer, processes and interprets that information, then operates its paper transport and image formation mechanisms to transcribe that information into a permanent form. Solid-state memory plays an important role in this operation. An EP printer operates on a fixed set of instructions that tells the printer how to operate. This program resides permanently inside the printer, so it is stored in a permanent memory device (a *ROM*—read-only memory). Other data, such as font styles and enhancements, also can be stored in permanent memory, or loaded into temporary memory (*RAM*—random-access memory) as needed.

Most information changes constantly as the printer works. Characters, graphics, and control codes received from the computer are only stored until they are processed. Main logic also requires locations to store control panel variables and results from calculations. Temporary memory devices are used to hold rapidly changing information during printer operation.

Permanent memory

As the name suggests, information in permanent memory is retained at all times, even while power is removed from its circuit. You might hear permanent memory referred to as *nonvolatile* or *read-only* memory. There are three basic classes of nonvolatile memory that you should be familiar with: *ROM* (read-only memory), *PROM* (programmable read-only memory), and *EPROM* (erasable programmable read-only memory). Your printer might use any of these classes, although the simpler PROM devices are encountered most frequently.

A ROM is the oldest and most straightforward class of permanent memory. Its information is specified by the purchaser, but the actual IC must be fabricated already programmed by the IC manufacturer. ROMs are rugged devices. Because their program is actually a physical part of the device, it can withstand a lot of electrical and physical abuse, yet still maintain its contents. However, once a ROM is programmed, its contents can never be altered. If a program change is needed, an entirely new device must be manufactured with any desired changes, then installed in the circuit. ROMs are often used in font cartridges that are manufactured in high volumes.

The PROM can be programmed by a printer manufacturer instead of relying on a ROM manufacturer to supply the programmed devices. A PROM can be programmed (or *burned*) once, but it can never be altered. Factory-fresh PROMS are built as a matrix of fusible links. An intact link produces a binary 0, and a burned link produces a binary 1. One link is available for each bit in the device. A special piece of equipment called a PROM Programmer is fed the desired data for each PROM address. It then steps through each PROM address and burns out any links where a binary 1 is desired. When the PROM is fully programmed, it contains the desired data or program. One or more PROMs are used to hold the printer internal program.

An EPROM can be erased and reprogrammed many times. Binary information is stored as electrical charges placed across *MOS* (metal-oxide semiconductor) transistors. One transistor is provided for each bit in the device. An absence of charge is a binary 0, and the presence of charge is a binary 1. Programming is very similar to that of a PROM. An EPROM programmer is loaded with the desired information for each location. It then steps through each location and locks charges into the appropriate bit locations. To erase an EPROM, you must remove charges from every bit location. That is accomplished by exposing the memory device (the die itself) to a source of short-wavelength ultraviolet light for a prescribed period of time. Light is introduced through the transparent quartz window on top of the IC package. You can tell an EPROM by the clear quartz window in the IC.

Temporary memory

Digital information contained in temporary memory can be altered or updated frequently, but it will only be retained as long as power is applied to the device. If power fails, all memory contents will be lost. This kind of memory device is referred to as *volatile*, *read-write*, or *RAM* (random-access memory. The term *random access* means the device can be accessed for reading or writing operations as needed. The two basic types of RAM that you should know are *static* and *dynamic*.

A static *RAM* (SRAM) uses conventional logic flip-flops (called *cells*) to store information. One cell is provided for each bit. A read/write control line is added to select between a read or write operation. During a write operation, any data bits existing on the data bus are loaded into the cells at the address specified on the address bus. If a read operation is selected, data contained at the selected address is made available to the data bus. Once data is loaded into a static RAM, it will remain until it is changed, or until power is removed. SRAMs are used heavily in printers.

DRAM (dynamic RAM) devices use small MOS cells to store data in the form of electrical charges. Although reading and writing operations remain virtually identical to those of SRAMs, DRAMs must be refreshed every few milliseconds, or their data will be lost. Refresh is provided by a combination of external circuitry and circuits within the DRAM chip itself. Although the need for refresh increases the complexity of a memory circuit, MOS technology offers very low power consumption and a large amount of storage space as compared to most SRAM devices.

Troubleshooting memory

Memory is usually one of the most reliable sections of an ECP, but when a failure does occur, the results can manifest themselves as lost characters, occasional operating hang-ups for no apparent reason, or up front initialization failure. The difficulty in testing memory is that it is virtually impossible to tell for sure just what location or bit the problem is coming from. To test a RAM device properly, a known pattern of data would have to be written to each location, then read back and compared to what was written. If there is a match, that location is assumed to be good. If there is no match, that location (and the entire IC) is defective. Most printers perform a memory check on initialization. Unfortunately, there is no way of performing this sort of test in the printer during normal operation.

ROM devices are almost as difficult to test. Each location would have to be read and a checksum value would have to be calculated based on the contents of each ROM address. The calculated checksum is compared against a checksum value on the ROM. If both checksum values match, the ROM is assumed to be valid. If there is no match, the ROM is defective. ROMs are tested with RAM when the printer is initialized.

As a result of these testing difficulties, there are no specific test procedures for memory circuits in particular. This book suggests the replacement of memory devices on a symptomatic basis only—when other symptoms point to the possibility of a memory failure. Although most memory problems will be detected and reported during the printer self-test process, memory can fail spontaneously during printer operation just like any other active component. Troubleshooting procedures in other sections of this book suggest where and when to suspect memory problems. When a memory problem is suspected, memory devices can be replaced systematically, but it also is acceptable to replace the ECP entirely.

Control panel

The printer control panel is the user interface between you and the printer. First, it allows you to operate certain immediate functions such as form feed, paper tray select, reset, or on/off line. Certain key combinations let you alter options and running modes (your user's manual will specify the exact key strokes and their effect). Indicators and LCD-alphanumeric displays also are included to display various printer status conditions or error codes. Figure 9-12 shows a simple block diagram for an EP printer control panel.

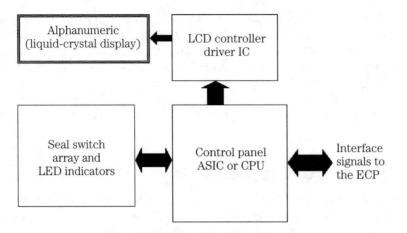

9-12 Block diagram of a typical control panel.

Sealed switches

Many printers use sealed membrane-type switches covered by a solid plastic strip containing the graphics for each key. A cross-sectional diagram of this arrangement is shown in Fig. 9-13. Membrane switches use a flexible metal diaphragm mounted in close proximity to a conductive base electrode at the switch bottom. Ordinarily, the diaphragm and base do not touch, so the switch is open. When you touch the proper location on a desired graphic, a solid plunger deforms the metal diaphragm and causes it to contact the base electrode. This action closes the switch. After you release the graphic, the metal diaphragm below returns to its original position and opens the switch again. The diaphragm design might snap a bit when pressed to provide you with a tactile sensation of positive contact.

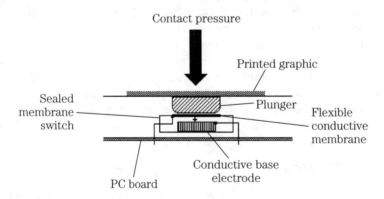

9-13 Diagram of a typical sealed (membrane) switch assembly.

Unfortunately, membrane switches are subject to breakdown with age, use, and environmental conditions. Although membrane switches are sealed to prevent disassembly, most are not hermetically sealed to keep out moisture and dust. Over time, oxidization can occur that prevents positive contact between the membrane and electrode. Regular use also can wear away at both contact surfaces and eventually cause bad or intermittent contacts—the switch might not always respond when you press it. Finally, regular use can cause the diaphragm to stretch or dislodge from its mounting, which can lead to a short circuit if the diaphragm fails to snap open when released.

Troubleshooting a control panel

Symptom 1 *The control panel does not function at all. No keys or indicators respond. Printer appears to operate normally under computer control.* Unplug the printer and allow 10 minutes for the unit to cool and discharge. Open your printer enclosure and expose the control panel circuit. Make sure that any connector(s) or wiring from the panel are installed properly and securely. If you have just finished reassembling the printer, perhaps you forgot to reconnect the control panel, or reconnected it improperly. Interconnecting wiring might have been crimped or broken during a previous repair.

If the printer appears to be working properly otherwise, the control panel module is probably defective and should be replaced. If a new control panel module does not resolve the problem (or the printer is acting erratically) the problem might be in the ECP. You can replace the ECP outright. If you have a schematic of the ECP, you can troubleshoot the control panel signals to the component level.

Symptom 2 *One or more keys is intermittent or defective. Excessive force or multiple attempts might be needed to operate the key(s). Printer appears to operate normally otherwise.* In almost every instance, this symptom is the result of faulty keys. Before replacing anything, check to make sure that all cables and wiring between the control panel and ECP are installed correctly and completely. If the cabling looks good, replace any questionable keys, or replace the entire control panel PC board assembly.

Symptom 3 *One or more indicators fail to function, or the LCD alphanumeric display appears erratic. The printer appears to operate normally otherwise.* Before attempting any troubleshooting or replacement, inspect any interconnecting cables or wiring between the control panel and ECP. Make sure that everything is connected correctly and completely. If the wiring is intact, the control panel has probably failed. Replace the control panel module. If the problem persists, the trouble is likely in the ECP itself. If you have schematics or technical information on your printer, you can troubleshoot the problem to the component level, otherwise, simply replace the ECP.

Main logic

Main-logic circuits are the heart and soul of your ECP. It typically includes a main microprocessor, one or more slave microprocessors (or ASICs), a clock oscillator, and any interconnecting "glue" logic components needed to tie these parts together. Your main-logic is responsible for directing all aspects of printer operation. Usually, it is possible to replace the ECP without a firm understanding of the circuitry itself, but this part of the chapter is intended to provide a more substantial background if you plan on working with ECPs in detail.

Microprocessor operations

If your EP printer could be compared to a symphony orchestra, the main microprocessor would be the conductor. Technically speaking, a *microprocessor* is a programmable logic device that can perform mathematical and logical manipulation of data, then produce desired output signals. All microprocessors are guided by a fixed series of instructions (called a *program*) that is stored in the printer permanent memory (ROM).

Although the microprocessors found in your printer are often less complex than those found in many computers, you can expect to find many of the same signals as shown in Fig. 9-14. Notice that a set of related signal wires (known as a *bus*) is often represented as a single wide line. This practice simplifies diagrams and schematics. Arrows are used to depict the possible flow of data. You will find three major busses: (1) the Address bus, (2) the Data bus, and (3) the Control bus.

Address lines specify the precise location of instructions or data anywhere in the printer. Address locations can refer to memory locations, operating addresses of slave microprocessors, ASICs, or other circuit-specific places. Because a microprocessor only generates address signals, it is always the controlling element in a main-logic circuit. The example microprocessor shown in Fig. 9-14 provides 16 address lines (A0 through A15) that can specify 2^{16} or 65536 unique locations. Your particular microprocessor probably offers more.

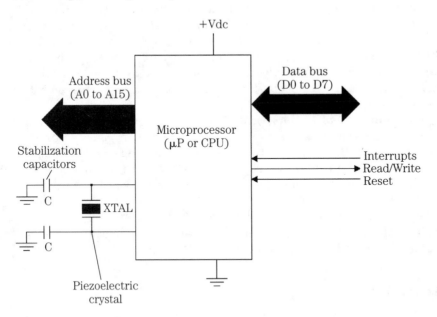

9-14 Major microprocessor signals.

A microprocessor can read data from or write data to any location specified by the address line. Data is sent over the *data bus*. For example, when the printer is first turned on, its microprocessor will automatically generate an address (usually 0000 hexidecimal) and attempts to read whatever character is available from that address. The microprocessor automatically assumes this to be its first instruction. If memory is defective, or the character at that address is somehow incorrect, the microprocessor will become hopelessly confused. This confusion leads to erratic or unpredictable operation.

During a write operation, the microprocessor will generate an address, then place valid data to that address. Any device that is active at that address (such as a RAM IC) will accept this data. Although a microprocessor can write to any address, not all addresses can accept data (such as ROM locations).

Finally, a microprocessor is equipped with a number of *control lines*. A read/write (R/W) signal informs the system whether the microprocessor is performing a read or a write operation at its selected address. An interrupt request (IRQ) is sent to the microprocessor whenever the printer must deal with an immediate problem or condition. There might be several levels or interrupts depending on the particular IC in use. Other control lines that you might encounter are the halt (HLT) and reset (RST) signals. Control signals are grouped together into the control bus.

The system clock

A microprocessor is a *sequential* device—instructions are executed one step at a time. As a result, timing becomes a very critical aspect of the microprocessor operation. Timing signals are provided by a precision oscillator built into the microprocessor itself. This oscillator is known as the *master clock* or *system clock*. To achieve a precise and stable clock, a piezoelectric crystal is added externally, and the oscillator will run at the resonant frequency of the crystal. For example, if an 8 MHz crystal were used, the clock would run at 8 MHz. Crystal frequencies are marked right on the part. You can read these clock pulses with your oscilloscope.

ASIC operations

A single microprocessor cannot possibly handle the wide variety of operations required of a typical printer. Older printer designs use a secondary (or *slave*) microprocessor to perform such tasks. The secondary microprocessor reduces processing time (or *overhead*) from the main microprocessor. Slave microprocessors have largely been replaced by Application Specific Integrated Circuits (ASICs). Where a slave microprocessor is somewhat generic in its structure and operation, an ASIC is a semi-custom or fully custom IC designed especially to serve specialized functions in a particular printer. An ASIC can generate all motor and laser/scanning unit control signals, operate the control panel, handle the printer serial and parallel interface, interpret sensor signals, and more. There might be one or more ASICs depending on the features and sophistication of your particular printer.

A key advantage to ASICs is that they eliminate a large portion of glue logic that results in smaller, simpler, and less power-demanding circuits. ASICs also work with minimum control from the main microprocessor, and are optimized for their use in a printer. ASICs can vary greatly in complexity depending on the intended capabilities of your printer, but most require large pin counts to provide enough signal lines. The disadvantage of ASICs is their potential difficulty to acquire as replacement parts. Manufacturers tightly control the production and distribution of these proprietary ICs.

Troubleshooting main logic

Most printer troubles do not involve a catastrophic failure of the main microprocessor or ASIC. Instead, typical ECP problems involve other components that directly interface with other parts of the printer such as the main motor, the laser/scanning assembly, the paper pickup and registration clutches, the control panel, and so on. However, there are some occasions when the microprocessor, system clock, or other highly integrated logic component will fail and cause serious problems for your printer.

When you first turn an EP printer on, main logic performs an initialization procedure that checks and clears all RAM locations, establishes a communication link with the host computer, and brings the high-voltage and fusing unit up to working levels. This procedure typically requires less than two minutes. If the printer does not initialize, or must be reinitialized during normal operation, the problem is almost certainly in the microprocessor or other main logic.

Symptom 1 *The printer does not initialize from a cold start.* No visible activity should take place in the printer after power is turned on, but power indicators are lit. Self-test does not work. If a printer encounters an error condition during its initialization, there will usually be some visible or audible indication of a fault condition. Audible tones, flashing light sequences, or an alphanumeric error codes are just some typical failure indicators. The user's manual for your particular printer will list any error codes and their meanings.

If the error is expected—that is, an error that is checked and handled by the printer software—the printer will simply wait until the error is corrected. Paper out errors are a commonly expected error. However, unexpected errors can cause the printer to freeze or behave erratically for no apparent reason. ROM or microprocessor defects are considered unexpected failures—main logic has no way of dealing with such problems—so there is no way of knowing just how your printer will respond (if it works at all).

Use your multimeter to measure the logic supply voltage levels in your printer (usually +5 Vdc). If this voltage is low or absent, logic devices will not function properly (if at all). A low or missing voltage output suggests a defective power supply. You can troubleshoot the dc supply using the procedures discussed in chapter 6, or simply replace the dc supply outright. If you do not have the tools or inclination to perform detailed procedures, replace the printer's ECP outright.

The next area to check is your system clock. If you have schematics or test instruments available, use your logic probe or oscilloscope to measure the oscillator signals on both sides of the piezoelectric crystal (make sure that your logic probe can measure frequencies as high as the rating of the crystal). If you are using an oscilloscope, there should be a roughly square wave at the frequency marked on the crystal. If the clock signal is low or missing, replace the crystal and stabilization capacitors. If this does not restore your clock source, replace the main microprocessor and retest the printer. If you do replace the microprocessor, be sure to install an appropriate IC socket first (if possible).

An initialization process can stall if the microprocessor detects a faulty RAM or ROM location. Unfortunately, there is no way to check a memory IC completely without using specialized test equipment. Some printers will display an error code indicating a memory error. If the error suggests a RAM fault, remove any option RAM boards or cards and retest the printer. If the problem disappears, that option module is defective. If the problem persists, replace the RAM chip(s) and retest the printer. If normal operation returns, you have isolated the defective component(s). Be sure to install appropriate IC holders (if possible) when replacing RAM chips. If a ROM error is indicated, replace the ROM IC(s) and retest the printer. If you do not have the schematics or test equipment for such detailed work, try replacing the ECP entirely.

If symptoms continue, replace the slave microprocessor or ASIC. A failure here can prevent motor operation, disable sensor signals, and cut off communication. The main microprocessor must interact closely with these components, so any fault here can hang up main microprocessor operation. As before, you might choose to simply replace the ECP outright if you do not have the technical data or test instruments to perform detailed checks.

Symptom 2 *Printer operation freezes or becomes highly erratic during operation.* You must activate the printer from a cold start to restore operation. Self-test might work until the printer freezes. Check the logic supply voltage with your multimeter. You should find about +5 Vdc. If this voltage is absent, low or intermittently low, logic devices will behave erratically. Troubleshoot the dc power supply using the procedures of chapter 6, or replace the dc supply outright.

A microprocessor requires constant access to its program ROM to operate properly. Each instruction and data location must be correct, or the main microprocessor will become hopelessly misdirected. If you find that the printer only operates to some consistent point where it freezes or acts strangely, the ROM might be defective. Replace the program ROM and retest the printer thoroughly. If normal operation returns, you have probably isolated the problem. Install an appropriate IC socket (if possible) before inserting a new ROM IC. If you do not have the schematics or test equipment to perform this type of procedure, try replacing the ECP outright.

Inspect all of your main-logic ICs for any devices that might appear excessively hot—especially if the printer has only been on for a short time. NEVER touch potentially hot components in a live circuit. Instead, smell around the circuit for unusually heated air, or hold the palm of your hand over the circuit. If a part seems unusually hot, spray it with liquid refrigerant. If normal operation returns temporarily (you might have to reactivate the printer or press a reset or on-line button), replace the thermally intermittent component. You also might replace the ECP entirely.

A
APPENDIX

Typical cleaning and maintenance

Whether your printer needs to be repaired or not, check and clean it periodically to ensure that no unexpected problems develop. Most laser printers have two replaceable components: the toner cartridge, and the development "engine" containing the EP drum and primary corona. These elements must be located and removed before cleaning can take place. On close inspection, you can probably find where these components are latched into a swinging deck. Release the latches and swing the toner cartridge and EP engine up and away from you.

Ordinary maintenance

Proper cleaning can be done in under 15 minutes, but you will need a supply of clean, lint-free cloths (preferably antistatic), a supply of clean, lint-free swabs, a few ounces of fresh denatured alcohol, and a supply of fresh, demineralized water. Avoid using ordinary tap water—minerals and contaminants can leave residue and cause corrosion.

Transfer corona

Use a clean, lint-free swab dipped lightly in fresh denatured alcohol to clean in and around the transfer corona rail. High-voltage potentials developed along the wire attracts dust and debris in the air nearby. Foreign matter collects on the corona wire and can eventually affect the image quality that is transferred to paper. Clean the length of wire gently but firmly. Be *extremely* careful *not* to break the corona wire. If it breaks, you must replace it. Also clean inside of the metal track surrounding the corona and be sure that all residue is removed.

Caution: Notice the thin monofilament line that is wrapped around the transfer corona case. This line prevents the attracted paper from being drawn into the corona case and jamming. You must be *extremely* careful to *avoid* breaking the line. If it breaks, the transfer corona will have to be re-wrapped or replaced.

Transfer guide

Next, use a clean, lint-free wipe dampened lightly with cold, demineralized water to clean the transfer guide area that is located just before the transfer corona assembly. The paper passes through this guide to be charged and get its image from the EP drum. Be sure to clean up any paper dust, debris, or residual toner that might be in the area. Do not use hot water to clean because hot water might melt the residual toner particles and cause a permanent mess. If it is possible to open the transfer guide lock tray, do so and clean inside the lock tray area. Also wipe the adjoining transfer guide roller. Use caution when cleaning—remember that the assemblies inside a laser printer are delicate and unforgiving. Excessive force and carelessness can easily result in collateral damage to the printer.

Static eliminator teeth

If you look between the transfer corona and the paper-feed guide, you will see a row of metallic teeth, which are the antistatic teeth. Once the paper (charged by the transfer corona) receives its latent toner image from the EP drum, the paper must be discharged. If discharge does not occur, the paper will retain a static charge, and each sheet will cling to one another. Worse, the charged sheet might jam in the printer paper path. Use a fine brush (such as a soft toothbrush or the brush end of the corona cleaning brush) to sweep away any accumulations of paper dust or debris.

Paper-feed guide

You should clean the paper feed guide next. Use a clean, lint-free wipe dampened lightly with cold, demineralized water to wipe away any dust or residual toner in the feed guide area. This is just straightforward cleaning, but be careful not to wipe debris into more critical areas of the printer.

Primary corona

Locate the primary corona integrated into the EP assembly that has been swung up and away from the rest of the printer. Like the transfer corona, the primary corona high-voltage operation attracts dust and debris from the air. As debris coats the wire, the corona surface charge becomes uneven and can affect the image quality on the EP drum. If you have not already found the primary corona cleaning brush that is usually kept near the paper-feed guide, look for it now. Gently ease the felt-tipped brush over the primary corona wire and slide the brush back and forth a few times to clean away any residue that might have accumulated. If you cannot find a brush, use a clean, lint-free swab dipped lightly in fresh denatured alcohol. Use caution when cleaning the primary corona. If the wire breaks, the entire EP cartridge will have to be replaced.

Separation pawls and cleaning pad

Clean the fuser separation pawls next. To find the pawls, open the printer fuser area located in the paper-ejection area (usually near the rear behind the EP drum). You will see several large plastic pawls (claws) leading to the fusing roller assembly. Use a clean, lint-free wipe dampened lightly with clean water to wipe the leading edge of each pawl. Be careful not to touch the fusing roller assembly.

Although the fusing rollers should not retain any toner, long-term use can wear the roller lubricant and allow spots of toner to remain on the roller. This residual toner can then appear on subsequent sheets of paper as spots or stains. A cleaning pad installed against the heated fusing roller wipes away any residual toner that might adhere during toner fusing, and helps to keep the fusing roller lubricated so toner will not stick. New cleaning pads are often included with new toner cartridges, so you need not usually replace the cleaning pad unless it is time to replace a toner cartridge.

Ozone filter

During normal operation, the coronas in your printer generate ozone from the high-energy ionization in surrounding air. Because ozone can be an irritant if there is enough exposure time, an ozone filter is often added to laser printers to reduce the amount of ozone gas released into the air. Typical ozone filters should be replaced after about 40,000 to 50,000 pages. Consider whether a new ozone filter is warranted. Most ozone filters are readily accessible while the printer is open for standard cleaning.

Reassemble the printer

Swing down the frame with the toner cartridge and EP assembly and relatch it into place. Unlatch and remove the toner cartridge itself and rock it back and forth along its long axis. Although many toner cartridge designs now allow for an agitator to keep toner evenly distributed, heavy or irregular use might allow the toner to become thin in one or more areas. Agitating the toner cartridge ensures that remaining toner will be evenly distributed and helps maximize toner cartridge life. Re-install the toner cartridge and secure all outer doors or panels. Cleaning should now be complete.

Restart the printer and initiate one or more self-tests. Compare the new and old self-tests. Under routine conditions, you should see no substantial difference between the tests. If the original self-test suggested maintenance neglect, you should see a noticeable improvement in the newer self-test.

Spotting maintenance-related problems

Incidental or infrequent users might overlook or neglect routine maintenance. When maintenance is neglected for too long, problems can occur in the final printed product. This part of the appendix shows you six typical maintenance-related problems.

Vertical fade

Vertical-fade problems are characterized by one or more faded streaks in the vertical direction. Under most circumstances, the toner level is getting low. Remove the toner cartridge and rock it back and forth to redistribute the toner evenly in its container. Replace the toner cartridge and adjust the print density.

Another cause of vertical fade is a dust- or debris-fouled transfer corona. Uneven charge distribution will allow light streaks to form. Check the transfer corona and clean it with a lint-free swab dipped lightly in fresh isopropyl alcohol. After cleaning the wire, swab out any residue within the corona case. Be extremely careful not to break the monofilament line wrapped around the corona. If the corona or wrap is broken, the transfer corona assembly will have to be replaced.

Dropouts

Dropouts generally appear as faded areas that are typically round in shape. The most common source of dropouts is due to problems with the paper itself. Paper with uneven moisture content or moist spots can result in dropouts. Even manufacturing defects in the paper can result in paper that produces dropout spots. Try fresh, dry paper from a different source. It also is remotely possible that the transfer corona is slightly dirty. If new paper fails to correct the problem, try cleaning the transfer corona as described earlier in this appendix.

Vertical lines

Vertical lines appear as one or more black streaks or smears directly from top to bottom of the page. When vertical lines appear, it usually indicates that one or more areas of the primary corona wire are fouled with accumulations of dust or debris. Clean the primary corona as described earlier in this appendix.

If the dark vertical marks appear smeared, it is possible that the fuser roller cleaning pad might be dirty and need replacement. If you do not have another cleaning pad on hand, you can gently clean the pad by removing it from the fuser assembly and gently brushing away any obvious accumulations of toner. A new cleaning pad should be installed as soon as possible. If a new cleaning pad is available, replace it according to the instructions for the particular printer.

Finally, it is possible that the EP drum has been scratched—this problem is rare, but can occur in printers serving a heavy work load. Unfortunately, the EP drum would have to be replaced, but the drum is typically part of the image formation engine, which can easily be replaced as an entire module. Replace the EP assembly and try the printer again.

Staining

A stain is typically a dark horizontal patch that reoccurs to lesser degrees down the page. Usually, the transport rollers that handle the paper are dirty. Clean the transport rollers and transport feed guide as described earlier in this appendix.

It also is possible that the fuser roller cleaning pad might be dirty and need replacement. If you do not have another cleaning pad on hand, you can gently clean the pad by removing it from the fuser assembly and gently brushing away any obvious accumulations of toner. Install a new cleaning pad as soon as possible. If a new cleaning pad is available, replace it according to the instructions for the particular printer.

Repetitive defects

Repetitive defects occur at regular intervals along the page. It is the spacing of the defects along the page that can really cue you to the actual problem. When the defects are spaced closely together (under 2 inches), there is possibly a problem with your paper transport roller(s). Check and clean each roller to remove all debris.

If the defects occur at intervals of 2 inches or greater, than there is a defect on either the development roller or the EP drum. Because both elements are incorporated into the image-formation engine (the EP cartridge), the entire assembly will have to be replaced in either case. Try replacing the EP cartridge.

Badly formed characters

Characters that are badly formed generally appear wavy or unsteady. This problem is almost always caused by paper stock that is too slick or slippery—the rollers have a difficult time handling the paper. Try the printer again with standard 20-pound xerography-grade paper. If the problem still occurs on standard paper, the printer scanner assembly is defective and the printer will require much more extensive bench repair.

Spotting user-related problems

A number of laser printer problems are caused by oversights or errors in judgment. This section of the appendix shows you some of the typical user-related problems found with laser printers.

1. *You are troubled with frequent jams.* A laser printer is only as good as the media that is used with it. Paper, envelopes, and labels must be chosen carefully to ensure proper printer operation. Remember that the printer mechanics are designed to work with media of certain thicknesses, textures, finishes, and weights. Media that is too heavy or flimsy or has an unusual or chemically treated finish might not be handled properly by the printer paper handling mechanism. Refer to the printer manual to find what is allowable, but standard white 20-pound bond xerography-grade paper should always work properly. The easiest solution here is to try different types of media.

2. *You find unusually light or dark images.* Adjust the print density control to achieve an optimum print density. Casual users often forget that they must adjust the print density wheel throughout the life of the toner cartridge. When a new toner cartridge is installed, the print density must be reduced or the resulting image might be too dark. As the toner is consumed, print will become lighter, so print density must be increased. Very light print might indicate that the toner cartridge must be replaced.

3. *The printer runs out of memory.* The memory in your printer serves as a buffer for data being sent from the host computer. For single-line printers, the buffer need not be very large. For laser printers, however, there must be enough memory to hold complex, high-resolution images (usually about 1 or 2MB). The entire image must be loaded into the laser printer buffer before

the page is printed. If the image to be printed is too large to fit into existing memory, the printer will register an OUT OF MEMORY error. Scale down the image or add memory to the printer.

4. *Paper holds too much static when ejected—could cause paper jams.* The paper itself is often to blame. Try a standard 20-pound xerographic-grade paper with average moisture content. Paper that is too dry retains static charges. It also is possible that the static eliminator teeth in the printer are not working properly. Try cleaning the static teeth as described earlier in this appendix.

B
APPENDIX

Comprehensive troubleshooting guides

Chapter 6: Troubleshooting ac and dc linear supplies

1. Power supply is completely dead. Laser printer does not operate and no power indicators are lit.
   a. Check ac line voltage powering the printer
   b. Check/replace the main ac fuse
   c. Check the outputs(s) from the supply
   d. Check the regulator(s)
   e. Check the filter(s)
   f. Check the rectifier(s)
   g. Check the transformer
   h. Check the supply PC board for damage
   i. Replace the power supply
2. Supply operation is intermittent. Printer operation cuts in and out along with the supply.
   a. Check ac line voltage powering the printer
   b. Check all power connectors and wiring
   c. Check the supply PC board for damage
   d. Check power components for thermal intermittents
   e. Replace the power supply
3. Laser printer is not operating properly. It might be functioning erratically or not at all. Power indicators might or might not be lit.
   a. Check ac line voltage powering the printer
   b. Check all power connectors and wiring
   c. Check the output(s) from the supply
   d. Check the regulator(s)
   e. Check the filter(s)

    f. Check the rectifier(s)
    g. Check the transformer
    h. Check the supply PC board for damage
    i. Replace the power supply
4. Fusing quality is intermittent or poor (toner smudges easily), or fuser fails to reach operating temperature within 60 to 120 seconds.
    a. Check all power connectors and wiring
    b. Check ac voltage powering the quartz heating lamp
    c. Check/replace the thermal cutout switch
    d. Check/replace the thermistor temperature sensor
    e. Replace the quartz heating lamp
    f. Replace the fusing assembly

Chapter 6: Switching-supply (dc) troubleshooting

1. Power supply is completely dead. Laser printer does not operate, and no power indicators are lit.
    a. Check ac line voltage powering the printer
    b. Check/replace the main ac fuse
    c. Check all power connectors and wiring
    d. Check the output(s) from the supply
    e. Check EMI shielding covering the supply
    f. Check primary and secondary ac at the transformer
    g. Check preswitched dc voltage levels
    h. Check chopped dc and filter(s)
    i. Check operation of the switching circuit or IC
    j. Replace the power supply
2. Supply operation is intermittent. Laser printer operation cuts in and out with the supply.
    a. Check ac line voltage powering the printer
    b. Check all power connectors and wiring
    c. Check the supply PC board for damage
    d. Check power components for thermal intermittents
    e. Replace the power supply
3. Laser printer is not operating properly. It might be functioning erratically or not at all. Power indicators might or might not be lit.
    a. Check ac line voltage powering the printer
    b. Check all power connectors and wiring
    c. Check the output(s) from the supply
    d. Check EMI shielding covering the supply
    e. Check primary and secondary ac at the transformer
    f. Check preswitched dc voltage levels
    g. Check chopped dc and filter(s)
    h. Check operation of the switching circuit or IC
    i. Replace the power supply

Chapter 6: High-voltage supply troubleshooting

1. Laser printing is too light or too dark.
   a. Check/adjust the contrast setting
   b. Replace the high-voltage supply
2. Cannot control laser printer contrast
   a. Check/adjust the contrast setting
   b. Replace the high-voltage supply

Chapter 7: System start-up problems

1. Nothing happens when power is turned on.
   a. Check all power connectors and wiring
   b. Check/replace the power supply
   c. Check/replace control panel cable
   d. Replace the ECP
   e. Replace the control panel
2. Your printer never leaves its warm-up mode. There is a continuous WARMING UP status code or message.
   a. Check the communication interface cable and host computer
   b. Check the control panel connectors and wiring
   c. Repair or replace the control panel
   d. Replace the ECP
3. You see a CHECKSUM ERROR message indicating a fault has been detected in the ECP program ROM.
   a. Try a cold restart of the printer
   b. Replace the ECP
4. You see an error indicating communication problems between the printer and computer.
   a. Check communication parameters (serial connection)
   b. Check cable and cable connections
   c. Check flow control settings
   d. Replace the ECP

Chapter 7: Laser delivery problems

1. You see a BEAM DETECTION error.
   a. Check/clean/replace the printer optics
   b. Check/replace the fiberoptic beam-detector cable
   c. Check/replace the laser or laser/scanning assembly
2. You see a BEAM LOST error.
   a. Check/replace the dc power supply
   b. Check/replace the fiberoptic beam detector cable
   c. Check mechanical safety interlock
   d. Replace the laser/scanning assembly

Chapter 7: Fusing assembly problems

1. You see a SERVICE error indicating a fusing malfunction.
    a. Check all connectors and wiring powering the fusing assembly
    b. Check all fusing circuit fuses or circuit breakers
    c. Check or replace the thermistor
    d. Check or replace the thermoprotector
    e. Check or replace the quartz lamp (or entire fusing assembly)
    f. Replace the ECP

Chapter 7: Image-formation problems

1. Pages are completely blacked out and might appear blotched with an undefined border.
    a. Check the primary corona in the EP cartridge/replace the EP cartridge
    b. Check/replace the fiberoptic beam-detector cable
    b. Check/replace the laser/scanning assembly
    c. Replace the ECP
2. Print is very faint.
    a. Check toner level/replace the EP cartridge
    b. Check paper quality
    c. Check/replace the transfer corona assembly
    d. Check/replace the high-voltage power supply
    e. Check/restore the drum ground integrity
3. Print appears speckled.
    a. Check/replace the fusing roller cleaning pad
    a. Check/replace the primary corona control grid or EP cartridge
    b. Replace the high-voltage power supply
4. There are one or more vertical white streaks in the print.
    a. Check toner level/replace the EP cartridge
    b. Check/clean the transfer corona assembly
    c. Check/replace the fiberoptic beam-detector cable
    d. Replace the laser/scanning assembly
5. Right-hand text appears missing or distorted.
    a. Check toner level/replace the EP cartridge
    b. Check/correct laser/scanning unit alignment and mounting
    c. Replace the laser/scanning assembly
6. You consistently encounter faulty image registration.
    a. Check paper quality
    b. Check the paper tray
    c. Check/replace the paper pickup assembly
    d. Check/replace the registration/transfer assembly
    e. Check/repair faulty drive train assembly
7. You encounter horizontal black lines spaced randomly through the print.
    a. Check/replace the laser beam detect sensor
    b. Check/replace the fiberoptic beam detector cable

    c. Check/replace the laser/scanner assembly

    d. Check all connectors and wiring from the scanner

  8. Print is just slightly faint.

    a. Check/adjust the contrast control setting

    b. Check paper quality

    c. Check toner level/replace the EP cartridge

    d. Check/replace the drum sensitivity sensor switches

    e. Check/clean/replace the transfer corona assembly

    f. Check or replace the high-voltage power supply

  9. Print has rough or suede appearance.

    a. Replace the ECP

10. Print appears smeared or fused improperly.

    a. Check/replace fusing roller cleaning pad(s)

    b. Clean/replace the fusing assembly

    c. Check/replace the static discharge comb

    d. Check the drive train

    e. Check and clear obstructions in the paper path

11. Printed images appear to be distorted.

    a. Check the paper path and mechanical assemblies

    b. Check or replace the scanning assembly

12. Print shows regular or repetitive defects.

    a. 3.75 inch defects—check or replace the EP cartridge

    b. 2 inch defects—check or replace the EP cartridge

    c. 3.060 inch defects—check or replace the fusing rollers

    d. Check and clean all other rollers or belts

13. The page appears completely black except for horizontal white stripes.

    a. Check/replace the fiberoptic beam detector cable

    b. Check/replace the laser/scanning assembly

14. The image appears skewed.

    a. Check the paper tray

    b. Check the paper

    c. Check/replace the paper pickup assembly

    d. Check/replace the registration assembly

    e. Check/clear any obstructions

15. The image is sized improperly.

    a. Check the paper tray tabs

    b. Check/replace the paper tray switches

    c. Replace the ECP

16. There are vertical black streaks in the image.

    a. Check/clean the primary corona

    b. Replace the EP cartridge

    c. Replace the ECP

Chapter 8: Paper problems

  1. You find a PAPER OUT message.

    a. Check the paper supply

    b. Check the paper tray and tray ID sensors
    c. Check the paper-out flag and sensor
    d. Replace the ECP
2. You see a PAPER JAM message.
    a. Check the paper supply quantity and quality
    b. Check/clear any obstructions in the paper path
    c. Check/replace the main motor
    d. Check/clear any obstructions in the gear train
    e. Check/replace the paper pickup assembly
    f. Check/replace the registration/transfer assembly
    g. Check/replace the fusing/exit assembly
3. The printed image appears with a smudged band and overprint.
    a. Check/replace the paper pickup assembly
    b. Check/clear any obstructions in the paper path

Chapter 8: Sensor and interlock problems

1. You see a PAPER OUT message even though paper is available.
    a. Check/replace the paper sensor mechanical flag
    b. Check/replace the paper sensor switch or optoisolator
    c. Replace the ECP
2. Fusing temperature control is ineffective. Temperature never climbs, or climbs out of control. This might affect print quality or initialization for EP printers.
    a. Check/replace the sensor thermistor
    b. Check/replace the thermoprotector
    c. Replace the ECP
3. You see a PRINTER OPEN message.
    a. Check that all printer housings are securely closed
    b. Check all housing sensor switches and actuators
    c. Check power to each housing sensor
    d. Replace the ECP

Chapter 8: Scanner motor/main motor problems

1. You see a general SCANNER ERROR message.
    a. Check all connectors and wiring between the laser/scanner assembly and ECP
    b. Check/replace the laser/scanner assembly
    c. Replace the ECP
2. The main motor does not turn, or turns intermittently.
    a. Check all connectors and wiring between the main motor and ECP
    b. Check/clear any obstructions in the paper path
    c. Check/replace the dc power supply
    d. Replace the main motor
    e. Replace the ECP

Chapter 8: EP cartridge problems

1. You see a NO EP CARTRIDGE message.
   a. Check/replace the EP cartridge
   b. Check/replace drum sensitivity sensor switches
   c. Replace the ECP
2. You see a TONER LOW message constantly, or the error never appears.
   a. Check/replace the EP cartridge
   b. Check/replace the high-voltage power supply assembly
   c. Replace the ECP

Chapter 9: Troubleshooting a parallel interface

1. Printer does not print at all. A PRINTER NOT READY error might occur at the computer. The printer self-test looks correct.
   a. Check/replace the communication interface cable
   b. Check printer DIP switch settings
   c. Check the handshake status lines
   d. Check data lines
   e. Check the Acknowledge pulse
   f. Check or replace the data latch IC or ASIC
   g. Replace the ECP

Chapter 9: Troubleshooting a serial interface

1. Printer does not operate at all. A PRINTER NOT READY error might occur at the printer. The printer self-test looks correct.
   a. Check/replace the communication interface cable
   b. Check printer DIP switch settings and parameters
   c. Check the hardware handshaking status lines(s)
   d. Check the flow control
   e. Check data and circuitry on the receive (Rx) line
   f. Check data and circuitry on the transmit (Tx) line
   g. Replace the ECP

Chapter 9: Troubleshooting a control panel

1. The control panel does not function at all. No keys or indicators respond. Printer appears to operate normally under computer control.
   a. Check all connectors or interconnecting wiring
   b. Check control panel supply voltage(s)
   c. Replace the control panel assembly
   d. Replace the ECP

2. One or more keys is intermittent or defective. Excessive force or multiple attempts might be needed to operate the key(s). Printer appears to operate normally otherwise.
   a. Check/replace all questionable keys
   b. Replace the control panel assembly
3. One or more indicators fail to function, or the LCD alphanumeric display appears erratic. Printer appears to operate normally otherwise.
   a. Check all connectors and interconnecting wiring
   b. Check control panel supply voltage(s)
   c. Replace the control panel assembly
   d. Replace the ECP

Chapter 9: Troubleshooting main logic

1. Printer does not initialize from a cold-start turn-on. There is no visible activity in the printer after power is turned on, but power indicators are lit. Self-test does not work.
   a. Check/replace the dc power supply
   b. Check/replace the system clock
   c. Replace printer RAM chips to check for defective memory
   d. Replace printer ASIC or slave microprocessor
   e. Replace printer program ROM
   f. Replace the ECP
2. Printer operation freezes or becomes highly erratic during operation. You must activate the printer from a cold-start to restore operation. Self-test might work until the printer freezes.
   a. Check/replace the dc power supply
   b. Replace printer program ROM
   c. Replace the printer main microprocessor
   d. Check for thermal intermittents
   e. Replace the ECP

C
APPENDIX

Vendors

The following index provides a comprehensive set of resources that you can use to help find parts, materials, and even outside service organizations. You can find this list updated regularly on TechNet BBS. This list should not be construed as an endorsement of the companies or services outlined below. You are advised to shop around and compare prices, warranties, and turnaround time before committing to any sale. Sales terms and conditions will vary between vendors. Use caution whenever dealing with mail-order organizations.

Component parts, data, and service

American Computer Repair, Inc.
  (printer module repair services)
6330 Farm Bureau Rd.
Allentown, PA 18106
Tel: 215-391-0100
Fax: 215-391-0431

Allied Electronics (general parts only)
7410 Pebble Dr.
Fort Worth, TX 76118
Tel: 800-433-5700

Computer Network Services (printer
  repair services)
100 Ford Rd.
Denville, NJ 07834
Tel: 201-625-4056
Fax: 201-625-9489

Digi-Key (general parts only)
701 Brooks Ave., S
P.O. Box 677
Thief River Falls, MN 56701-0677
Tel: 800-344-4539
Fax: 218-681-3380

Electroservice Labs (full-service repair
  center)
6085 Sikorsky St.
Ventura, CA 93003
Tel: 805-644-2944
Fax: 805-644-5006

Impact (laser fuser assembly repair)
10435 Burnet Rd.
Suite 114
Austin, TX 78758
Tel: 512-832-9151
Fax: 512-832-9321

Kennsco Component Services (repair
  services)
2500 Broadway St., NE
Minneapolis, MN 55413
Tel: 800-525-5608
Fax: 612-623-4489

MicroMedics (complete printer parts
  and repair services)
6625 W. Jarvis
Niles, IL 60714
Tel: 800-678-5300
Tel: 708-647-1010

Mouser Electronics (general parts only)
2401 Highway 287 N.
Mansfield, TX 76063-4827
Tel: 800-346-6873
Tel: 817-483-4422
Fax: 817-483-0931

National Parts Depot, Inc. (printer parts
  and technical manuals)
31 Elkay Dr.
Chester, NY 10918
Tel: 914-469-4800
Fax: 914-469-4855

Northstar (printer parts and service)
7940 Ranchers Rd.
Minneapolis, MN 55432
Tel: 800-969-0009
Fax: 612-785-1135

ProAmerica (printer parts)
650 International Pkwy.
Suite 180
Richardson, TX 75081
Tel: 800-888-9600
Tel: 214-680-9600
Fax: 214-690-8648

Robec Spare Parts (printer parts and
  technical manuals)
425 Privet Rd.
Horsham, PA 19044
Tel: 800-223-7078
Tel: 215-675-9300
Fax: 215-672-5605

The Printer Works (printer parts, service,
  supplies, accessories, and service)
3481 Arden Rd.
Hayward, CA 94545
Tel: 800-235-6116
Tel: 510-887-6116
Fax: 510-786-0589

Unicomp, Inc. (printer parts and repairs)
2400 W. Fifth St.
Santa Ana, CA 92703
Tel: 800-359-5092
Tel: 714-571-1900
Fax: 714-571-1909

General reference

B+K Precision (a division of Maxtec
  International)
6470 W. Cortland St.
Chicago, IL 60635
Tel: 312-889-1448
Fax: 312-794-9740

Dynamic Learning Systems
Attn: Stephen J. Bigelow
P.O. Box 805
Marlboro, MA 01752
TechNet BBS: 508-366-7683 (N/8/1)

[Professional BBS subscriptions and
The PC Toolbox newsletter $39
per year U.S. See order form in the back
of the book.]

Howard W. Sams & Co.
2647 Waterfront Parkway East Dr.
Indianapolis, IN 46214-2041
Tel: 800-428-7267

Laser printer manufacturers

Brother International Corp.
200 Cottontail Ln.
Somerset, NJ 08875-6714
Tel: 800-276-7746
Tel: 908-356-8880

Citizen America Corp.
2450 Broadway
P.O. Box 4003
Santa Monica, CA 90411
Tel: 310-453-0614

Digital Equipment Corp.
146 Main St.
Maynard, MA 01754-2571
Tel: 800-344-4825
Tel: 508-493-5111

Epson America, Inc.
20770 Madrona Ave.
Torrance, CA 90509-2842
Tel: 800-289-3776
Tel: 310-782-0770

Hewlett-Packard Co.
P.O. Box 58059
MS511L-SJ
Santa Clara, CA 95051-8059
Tel: 800-752-0900

LaserMaster Corp.
6900 Shady Oak Rd.
Eden Prairie, MN 55344
Tel: 800-950-6868
Tel: 612-944-9330

Lexmark International, Inc.
740 New Circle Rd.
Lexington, KY 40511-1847
Tel: 800-358-5835
Tel: 606-232-2000

NEC Technologies, Inc.
1414 Massachusetts Ave.
Boxboro, MA 01719
Tel: 800-388-8888
Tel: 508-264-8000

NewGen Systems Corp.
17550 Newhope St.
Fountain Valley, CA 92708
Tel: 800-756-0556
Tel: 714-641-2800

Okidata
532 Fellowship Rd.
Mt. Laurel, NJ 08054
Tel: 800-654-3282
Tel: 609-235-2600

Panasonic Communications &
  Systems Co.
Two Panasonic Way
Secaucus, NJ 07094
Tel: 800-742-8086

QMS, Inc.
One Magnum Pass
Mobile, AL 36689-1250
Tel: 205-639-4400

Samsung Electronics America, Inc.
Information Systems Division
105 Challenger Rd.
Ridgefield Park, NJ 07660
Tel: 800-446-0262
Tel: 201-229-4000

Sharp Electronics Corp.
Sharp Plaza
Mahwah, NJ 07430
Tel: 800-526-0522
Tel: 201-529-9593

Star Micronics America, Inc.
420 Lexington Ave., #2702
New York, NY 10170
Tel: 800-447-4700
Tel: 212-986-6770

Tandy Corp.
1500 One Tandy Center
Fort Worth, TX 76102
Tel: 817-390-3011

Texas Instruments, Inc.
P.O. Box 200230
Austin, TX 78720
Tel: 800-527-3500
Tel: 817-771-5856

Xante Corp.
2559 Emogene St.
Mobile, AL 36606
Tel: 800-926-8839
Tel: 205-476-8189

Laser printer supplies and materials

Global Computer Supplies
1 Harbor Park Dr.
Department 52
Port Washington, NY 11050
Tel: 800-845-6225
Tel: 516-625-6200
Fax: 516-625-6683

Misco
One Misco Plaza
Holmdel, NJ 07733
Tel: 800-876-4726
Fax: 908-264-5955

Glossary

ACK (acknowledge) A handshaking signal sent from printer to computer indicating that the printer has successfully received a character.

anode The positive electrode of a two-terminal electronic device.

ASAP (Advanced Systems Architecture for PostScript) A printer controller design that has been trademarked by QMS

ASCII (American Standard Code for Information Interchange) A standard set of binary codes that define basic letters, numbers, and symbols.

ASIC (application-specific IC) A specialized IC developed to serve a specific function (or set of functions) in a printer.

base One of three electrodes on a bipolar transistor.

baud rate The rate of serial data transmissions which is measured in *PBS* (bits per second).

binary A number system consisting of only two digits.

bitmap A two-dimensional array of dots that compose an image.

BITBLT (BIT block-level transfer) The transfer of part of a bitmap image from one area in the laser printer memory to another.

BUSY A handshaking signal sent from printer to computer indicating that the printer cannot accept any more characters.

capacitance The measure of a device's ability to store an electric charge, measured in farads, microfarads, or picofarads.

capacitor A device used to store an electrical charge.

cathode The negative electrode of a two-terminal electronic device.

clutch A mechanism used to switch mechanical force into or out of a mechanical assembly.

collector One of three electrodes on a bipolar transistor.

continuity The integrity of a connection measured as a very low resistance by an ohmmeter.

corona A field of concentrated electrical charge produced by a large voltage potential. Corona wires form one electrode of this voltage potential. There are two coronas in a laser printer; the primary corona, and the transfer corona.

CPI (characters per inch) The number of characters that will fit onto one inch of horizontal line space, also called character pitch.

CPU (central processing unit) The major controlling logic element in your printer's circuitry.

CTS (clear to send) A serial handshaking line at the computer usually connected to the RTS line of a printer.

data Any of eight parallel data lines that carry binary information from computer to printer.

data buffer Temporary memory where characters from the computer are stored by the printer prior to printing.

DCD (Data Carrier Detect) A serial handshaking line usually found in serial modem interfaces.

developing The movement of toner from a toner supply to the latent image written to the charged drum surface.

diode A two-terminal electronic device used to conduct current in one direction only.

dpi (dots per inch) A laser printer's resolution expressed as the number of individually addressable dots that can be placed in both the horizontal and vertical directions.

driver An amplifier used to convert low-power signals into high-power signals.

DSR (Data Set Ready) The primary computer signal line for hardware handshaking over a serial interface. It is connected to the DTR line at the printer.

DTR (Data Terminal Ready) The primary serial printer signal for hardware handshaking over a serial interface. It is connected to the DSR pin at the computer.

ECP (electronic control package) A generic term referring to the electronic assembly used to control a laser printer. The ECP consists of main logic, memory, drivers, and a control panel. Also called a *controller*.

emitter One of three electrodes on a bipolar transistor.

EP (electrophotographic) Also called *electrostatic* (see ES).

EPROM (erasable programmable read-only memory) An advanced type of permanent memory that can be erased and re-written to many times.

ES (electrostatic) A process of creating images using the forces of high-voltage to attract or repel media (toner) as needed to form the desired image.

exposure The process of discharging the drum after cleaning to remove any electrical charges on the drum photosensitive surface.

fixing See **fusing**.

font A character set of particular size, style, and spacing.

font scaling The process of producing bitmaps of various sizes from a generic source of size-independent character information.

fusing The process of using heat and pressure to bond toner to a porous paper surface.

gates Integrated circuits used to perform simple logical operations on binary data in digital systems.

GND (ground) A common electrical reference point for electronic data signals.

GPIB (general-purpose interface bus) A parallel communication interface intended primarily for networked instrumentation, also known as IEEE 488.

gray scale A series of shades running from white to black. For laser printers, shades are produced by creating various patterns of dots (called *dithering*).

inductance The measure of the ability of a device to store a magnetic charge, measured in henrys, millihenrys, or microhenrys.

inductor A device used to store a magnetic charge.

initialization Restoring default or start-up conditions to the printer due to fault or power-up.

landscape The orientation of characters or images on a page that runs the short way (on an 8.5 × 11 inch page, the 8.5 inch sides would be vertical).

laser A device producing a narrow intense beam of coherent, single-wavelength light waves.

LCD (liquid crystal display) A display using character images formed from layers of voltage-polarized liquid. In its off state, the liquid is clear. In its on state, the liquid is opaque.

LED (light-emitting diode) A semiconductor device designed such that photons of light are liberated when its p-n junction is forward biased.

lpi (lines per inch) The number of horizontal lines that fit into one inch of vertical page space, also known as line pitch.

MB (Megabyte) An amount of memory or storage area. Each megabyte is 1,048,576 bytes.

microprocessor A complex programmable logic device that will perform various logical operations and calculations based on predetermined program instructions.

motor An electromechanical device used to convert electrical energy into mechanical motion. There are several types of motors used in laser printers.

MTBF (mean time between failures) A measure of a device's reliability expressed as time or an amount of use.

multimeter A versatile test instrument used to test such circuit parameters as voltage, current, and resistance. Also called a DVM or VOM.

parity An extra bit added to a serial data word used to check for errors in communication.

pawl Curved plastic assemblies resembling claws that guide charged paper through the printer.

PCL (printer command language) A popular printer control language developed by Hewlett-Packard that is used or emulated by almost all 300 × 300 dpi resolution laser printers.

PDL (page-description language) A resolution-independent printer language that describes the elements of a printed page. Commonly used with PostScript or PostScript-compatible printers.

PE (Paper Error) A handshaking signal sent from the printer to tell the computer that paper is exhausted.

photosensitive A material or device that reacts electrically when exposed to light.

piezoelectric The property of certain materials to vibrate when voltage is applied to them.

portrait The orientation of characters or images on a page that runs the long way (on an 8.5 × 11 inch page, the 11 inch sides would run vertical).

ppm (pages per minute) The maximum speed at which a laser printer engine can move paper.

RAM (random-access memory) A temporary memory device used to store digital information.

regulator An electronic device used to control the output of voltage and current from a power supply.

resistance The measure of a device's ability to limit electrical current, measured in ohms, kilohms, or megohms.

resistor A device used to limit the flow of electrical current.

RET (resolution enhancement technology) Introduced by Hewlett-Packard for the LaserJet III, RET improves edge definition by varying the size of dots around the edges of bitmaps.

ROM (read-only memory) A permanent memory device used to store digital information.

RTS (Request To Send) A printer serial handshaking line usually connected to the CTS line of a computer.

Rx (Receive Data) This is the serial input line. The printer Rx line is connected to the computer Tx line.

scanner In laser printers, the scanner assembly uses a rotating hexagonal mirror to direct the writing laser beam along the photosensitive drum surface. Dots are formed by turning the laser beam on or off while the beam is being scanned.

Select A control signal from the computer that prepares the printer to receive data.

separation pad A soft rubber pad in the paper transfer assembly that prevents more than one page at a time from entering the printer.

soft font Font vector or bitmap data on diskette or other computer media (such as CD-ROM).

solenoid An electromechanical device consisting of a coil of wire wrapped around a core which is free to move.

Strobe A handshaking line from the computer that tells the printer to accept valid parallel data on its data lines.

thermistor A temperature sensing device used to regulate temperature in the fusing roller assembly.

toner A fine powder of plastic, iron, and pigments used to form images in electrostatic printing systems.

transfer The process of attracting the developed image off the drum and onto the charged paper surface.

transistor A three-terminal electronic device whose output signal is proportional to its input signal. A transistor can act as an amplifier or a switch.

transformer A device used to step the voltage and current levels of ac signals.

Tx (Transmit Data) This is the data output line for serial devices. The computer Tx line is connected to the printer Rx line.

Index

Illustrations are in **boldface.**

About the Author

Stephen J. Bigelow is an electrical engineer specializing in the repair of computer and telecommunications equipment. His articles frequently appear in *Radio Electronics*, *Popular Electronics*, and *Modern Electronics*. He is the author of several McGraw-Hill books, including *Maintain and Repair Your Computer Printer and Save a Bundle*, *Maintain and Repair Your Notebook, Palmtop, or Pen Computer*, *Troubleshooting and Repairing Computer Printers*, and *Troubleshooting and Repairing PC Drives and Memory Systems*.

Tired of fixing your PC "in the dark"?

Now you don't have to! Subscribe to:

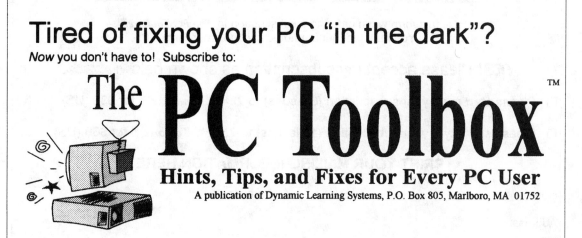

The PC Toolbox™

Hints, Tips, and Fixes for Every PC User

A publication of Dynamic Learning Systems, P.O. Box 805, Marlboro, MA 01752

Finally, there is a newsletter that brings computer technology into perspective for PC users and electronics enthusiasts. Each issue of *The PC Toolbox*™ is packed with valuable information that you can use to keep your system running better and longer:

♦ Step-by-step PC troubleshooting information that is easy to read and inexpensive to follow. Troubleshooting articles cover all aspects of the PC and its peripherals including monitors, modems, and printers.

♦ Learn how to upgrade and configure your PC with hard drives, video boards, memory, OverDrive™ microprocessors. scanners, and more.

♦ Learn to understand the latest terminology and computer concepts. Shop for new and used systems with confidence.

♦ Improve system performance by optimizing your MS-DOS™ and Windows™ operating systems.

♦ Review the latest PC, DOS, and Windows™ books and software *before* you buy them.

♦ Get access to over 1500 PC utilities and diagnostic shareware programs through the Dynamic Learning Systems BBS.

Each issue is 16 pages long, and a 1 year subscription is 6 issues. There is no advertising space and no hype - just practical tips and techniques cover to cover that you can put to work right away. If you use PCs or repair them, try a subscription to *The PC Toolbox*™. If you are not completely satisfied within 90 days, just cancel your subscription and receive a full refund - *no questions asked!* See the accompanying order form, or contact: **Dynamic Learning Systems, P.O. Box 805, Marlboro, MA 01752**.

MS-DOS and Windows are trademarks of Microsoft Corporation. The PC Toolbox is a trademark of Dynamic Learning Systems. The PC Toolbox logo is Copyright©1994 Corel Corporation. Used under license.

A Risk-Free Subscription Offer

Use this form when ordering a subscription to *The PC Toolbox*™.
You may tear out or photocopy this order form.

YES! Please accept my subscription as shown below: (check any one)

☐ Please start my 1 year subscription (6 issues) to *The PC Toolbox*™ for $39 (US).

☐ Please start my 2 year subscription (12 issues) to *The PC Toolbox*™ for $69 (US).

PRINT YOUR MAILING INFORMATION HERE:

Name: Company:

Address:

City, State, Zip+4:

Country:

Telephone: (   )   Fax: (   )

PLACING YOUR ORDER: (have your MasterCard or VISA ready)

By FAX: FAX this completed order form 24 hours/day, 7 days/week to **508-898-9995**

By PHONE: Phone in your order (Monday to Friday, 9am to 4pm EST) to **508-366-9487**

By BBS: Place your order on-line 24 hours/day, 7 days/week at **508-366-7683**

☐ MasterCard   Card#:_ _ _ _ _ _ _ _ _ _ _ _ _ _ _ Signature:_____

☐ VISA   Exp._____

OR By **MAIL:**   Mail your order form along with a check or money order to
Dynamic Learning Systems, P.O. Box 805, Marlboro, MA 01752

Is there anyone else who should be receiving *The PC Toolbox*™?

Name: Address:

City: State: Zip:

Name: Address:

City: State: Zip: